Southwestern
TREES

A Guide to the Trees of Arizona, New Mexico, and the Southwestern United States

STEVE W. CHADDE

AN ORCHARD INNOVATIONS FIELD GUIDE

GAMBEL OAK

SOUTHWESTERN TREES
A Guide to the Trees of Arizona, New Mexico, and the Southwestern United States

STEVE W. CHADDE

An Orchard Innovations Field Guide
ISBN: 978-1-951682-16-3

The author can be reached at: *steve@orchardinnovations.com*
VERSION 2, DECEMBER 28, 2024

Contents

Introduction · 5

Conspectus: Plant Families & Genera · 9

Tree Descriptions · 17

Conifers · 17

Broadleaf Trees · 64

Agaves and Cacti · 231

Palms · 253

Glossary · 255

Ecoregion Map – Arizona · 260

Ecoregion Map – New Mexico · 261

Acknowledgments · 262

Online Resources · 262

Index of Scientific Names · 263

Index of Common Names · 267

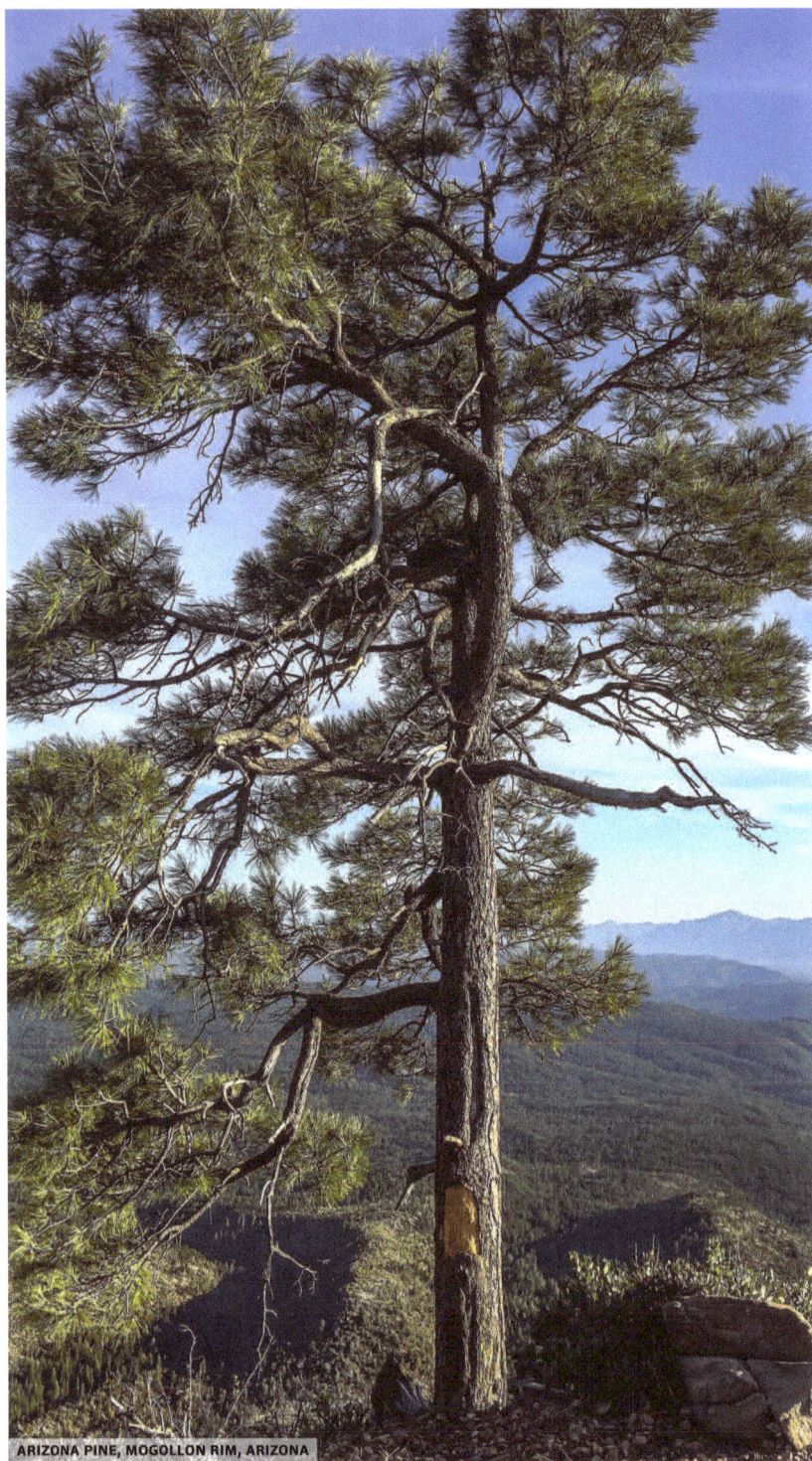

ARIZONA PINE, MOGOLLON RIM, ARIZONA

Introduction

THE SOUTHWEST, where the low, hot, barren Mexican deserts meet the lofty, cool, forested Rocky Mountains in New Mexico and Arizona, has an unsuspected richness of native trees. The deserts are not really treeless but contain here and there, scattered along the drainages and rocky slopes, strange spiny dwarf trees, such as yuccas, paloverdes, and cacti, found nowhere else in the United States. On foothills and mesas are broad zones of orchard-like woodlands of pinyons, junipers, and evergreen oaks. At higher elevations are mountain ranges and high plateaus, covered with valuable forests of tall pines, spruces, and firs. Within these extensive timberlands of New Mexico and Arizona are located thirteen national forests and the largest area of continuous ponderosa pine forests in the West.

Southwestern Trees is a field guide to the trees found in Arizona and New Mexico, extending to the surrounding areas of southeastern California, southern Nevada, southern Utah, southern Colorado, western Texas, and northern Mexico. Included are descriptions, color photographs, and distribution maps for the coniferous and broadleaf trees of the region, as well as the sometimes tree-like species of agaves and yuccas, so characteristic of the region's desert flora. Also included are a number of species more commonly found as shrubs, but occasionally reaching tree-size in favorable environments.

Within each group, trees are arranged alphabetically by plant family and genus. If more than one species occurs within a family or genus, a key is provided. By using a combination of keys,

plant and habitat descriptions, distribution maps, and illustrations, you should be able to accurately identify any tree encountered in the region. For more information about the region's trees, refer to the list of online resources on page 262.

Of course, the division between trees and shrubs is gradual, especially in the desert, and the number of species recognized as trees depends upon the definition accepted. Here, trees are considered as woody plants having a single, well-defined, perennial stem or trunk at least 2 inches (5 cm) in diameter at breast height, a height of at least 10 feet (3 m), and a somewhat definitely formed crown of foliage. Tree sizes refer to trees growing in the Southwest, and trunk diameters are measured at breast height (4½ feet).

Taxonomy

Existing tree guides for the region are mostly old, and the scientific names of many plants (and sometimes their family assignments) have changed. Names have been updated in this guide, in most cases following that of the published volumes of *The Flora of North America* series (1993+), the *Synthesis of the North American Flora* (Kartesz 2014), and the *Integrated Taxonomic Information System* (ITIS, *www.itis.gov*). Previously used named (synonyms) are provided as needed.

A note on the maps

Where available, two maps are provided for each tree. The map on the lower left of the page is based on the *National Individual Tree Species Atlas* (USDA Forest Service 2015). These maps are based on modeled (shown in purple) as well as actual mapped tree distributions (shown in pink), overlaid on a map showing forested areas (green). An intermediate color is used when modeled and mapped distributions overlap one another.

The lower right map shows known county occurrences (bright green) based on data provided by the Biota of North America Program (BONAP). Dark green indicates presence in a state; cyan and blue indicate the species is present but introduced from elsewhere (i.e., not native to the state); yellow indicates the species is uncommon or rare. Finally, a blue-green color (as in honey-locust) indicates the species is considered adventive (i.e., not truly native in the state) from another location in the USA.

Scope

The guide includes descriptions for 143 species of trees; additionally, a number of minor trees, and shrubs rarely reaching tree size are also noted.

These southwestern trees are classified into 66 genera and 33 plant families. Families with greatest numbers of tree species in the region are: pea family, 18 species; beech family (the oaks), 17; pine family, 15; willow family, 14; and rose family, 12. The largest tree genera here are: oaks (*Quercus,* 16 species) and willows (*Salix,* 11 species); pines (*Pinus*), 10 species; junipers (*Juniperus*), 7 species; ashes (*Fraxinus*) and yuccas (*Yucca*), 5 species each; cottonwoods (*Populus*), 4 species; and cherries (*Prunus*) and chollas (*Cylindropuntia*), 4 species each.

Geographic distribution

Based on geographic distribution, southwestern trees may be arranged in several groups. Most coniferous and other trees of the high mountain forests, such as ponderosa pine, Douglas-fir, and aspen, are widespread in the Rocky

Mountains and extend southward to the Southwest. Another and more prominent group of desert trees and other trees, chiefly of lower elevations, is southwestern and Mexican, from southwestern Texas and northern Mexico northward into southern New Mexico and central Arizona, or beyond to southwestern Utah, southern Nevada, and southeastern California. Gray oak, catclaw acacia, and desert-willow are examples. An interesting division includes the trees of the Mexican border region which barely reach mountains in the extreme southwestern corner of New Mexico (Hidalgo County) and the southeastern corner of Arizona (Cochise, Santa Cruz, and Pima Counties). Examples are Apache pine, Mexican blue oak, and Arizona madrone.

A very small group of trees of the short-grass plains, such as plains cottonwood and little walnut, enter eastern New Mexico from the east. A few shrubby tree species of the California chaparral, such as sugar sumac, hollyleaf buckthorn, and California fremontia, are present also in the chaparral zone of central Arizona. Some trees of middle elevations, such as pinyon, one-seed juniper, and velvet ash, have their ranges centering in these two States.

Desert trees chiefly are limited to one or more of the four southwestern desert regions characterized by creosotebush and mesquite: (1) The Chihuahuan desert in southern New Mexico, mostly in Tularosa Basin and Rio Grande basin (Torrey yucca); (2) the Arizona desert of southern and central Arizona (saguaro); (3) the Colorado desert of southwestern Arizona (smokethorn); and (4) the Mojave desert of northwestern Arizona (Joshua-tree).

Vegetation of Arizona and New Mexico

Most species of southwestern trees have a geographic and altitudinal distribution which follows closely the extent of one or more zones or types of natural vegetation.

The distribution of plant life, or vegetation, over the earth's surface is dependent upon the interaction of various external factors, such as climate (temperature and rainfall), topography and soil, other living things (including humans), and fire. Forests occur where there is a long, warm, moist growing season and where the soil is moist throughout the year. Grasslands are found generally where most of the rainfall is in the growing season and where low moisture or a dry season prevents tree growth. Deserts, with their sparse vegetation, are warm or hot regions where moisture is insufficient to support dense growths of trees or grass.

Within the broad climatic zones (tropical, temperate, and frigid), topography is important. A difference in elevation (altitude) of 1,000 feet affects climate and distribution of plant and animal life in about the same way as a distance north or south (latitude) of 300 miles at sea level. Thus, the summit of a mountain peak a mile (5,280 feet) above its base may have vegetation like that at low elevations about 1,600 miles northward.

In the Southwest, variations in climate and topography are extreme. Elevations in New Mexico range from 2,876 feet above sea level where the Pecos River leaves the southern border in the desert zone to 13,306 feet on the cold, barren, alpine summit of North Truchas Peak in the Sangre de Cristo Range. The southwestern corner of Arizona, elevation only 100 feet, has subtropical desert plant life, while San Francisco Mountain, elevation 12,655 feet, has a timberline and tundra vegetation.

The principal types of natural vegetation in Arizona and New Mexico are summarized on page 8. Ecoregion maps for each state are provided on pages 260 and 261. Ecoregions provide a broad overview of the primary ecosystems found in each state (more information on ecoregions is available at *www.epa.gov/eco-research*).

A representative southwestern mountain peak may have the following zones or belts of vegetation on its slopes (from low to high): *Grassland, pinyon-juniper woodland, ponderosa pine forest, Douglas-fir forest, spruce-fir forest,* and *alpine tundra.* These zones, of course, are not sharply limited but blend into one another. Rainfall generally increases with elevation. Vast plains lack water and are treeless, while the higher mountains rising upward from grassland or desert have a lower timberline where trees can successfully establish, and are forested above.

The elevation of the different zones naturally increases slightly from the northern to southern parts of these large States. On a mountain, the lower zones extend highest on the relatively warmer and drier southern and southwestern exposures, while the higher zones extend lowest on northern and northeastern slopes and in cool, shaded canyons. Zones are lower where rainfall is high and rise on plateaus or similar broad uplands.

Principal vegetation types, Arizona & New Mexico

ALPINE TUNDRA mountain avens, alpine sedges, alpine grasses
Elevation (feet) 11,500-13,306
Annual rainfall (inches) 30-35
Location in New Mexico Above timberline of summits of Sangre de Cristo Range and Jemez Mountains in northern part.
Location in Arizona Above timberline on summit of San Francisco Mountain.

SPRUCE-FIR FOREST Engelmann spruce, subalpine fir, corkbark fir
Elevation (feet) 8,500-12,000
Annual rainfall (inches) 30-35
Location in New Mexico High mountains, especially Sangre de Cristo Range, Jemez, Sacramento, and Mogollon Mountains.
Location in Arizona High mountains, especially White Mountains, San Francisco Mountain, and Kaibab Plateau.

DOUGLAS-FIR FOREST Douglas-fir, white fir, quaking aspen, limber pine
Elevation (feet) 8,000-9,500
Annual rainfall (inches) 25-30
Location in New Mexico High mountains in western two-thirds.
Location in Arizona High mountains in e and n parts.

PONDEROSA PINE FOREST ponderosa pine, Arizona pine
Elevation (feet) 5,500-8,500
Annual rainfall (inches) 19-25
Location in New Mexico Mountains in western two-thirds and northeastern corner.
Location in Arizona Mountains and plateaus in ne half.

PINYON-JUNIPER WOODLAND pinyon, Utah juniper, one-seed juniper, alligator juniper, Rocky Mountain juniper
Elevation (feet) 4,500-7,500
Annual rainfall (inches) 12-20
Location in New Mexico Plateaus, foothills, and mountains except e quarter.
Location in Arizona Plateaus and mountains in ne half.

OAK WOODLAND Emory oak, gray oak, Mexican blue oak, Arizona white oak
Elevation (feet) 4,500-6,000

Annual rainfall (inches) 12-20
Location in New Mexico Foothills and mountains in s quarter.
Location in Arizona Foothills and mountains in se and central parts.

CHAPARRAL shrub live oak, manzanitas, sumacs, cliffrose, ceanothuses
Elevation (feet) 4,000-5,500
Annual rainfall (inches) 13-25
Location in New Mexico none.
Location in Arizona Mountains in central part.

OAK BRUSH Havard's oak
Elevation (feet) 3,600-4,800
Annual rainfall (inches) 13-18
Location in New Mexico se corner
Location in Arizona none.

SHORT GRASS blue grama, hairy grama, galleta, buffalograss
Elevation (feet) 4,500-6,500
Annual rainfall (inches) 9-20
Location in New Mexico Plains, mostly in e, central, and n parts.
Location in Arizona Plains and plateaus in n part.

DESERT GRASS black grama, tobosa, dropseeds
Elevation (feet) 3,000-5,000
Annual rainfall (inches) 9-18
Location in New Mexico Plains in sw half.
Location in Arizona Plains in se part.

SAGEBRUSH big sagebrush, blackbrush
Elevation (feet) 2,500-6,000
Annual rainfall (inches) 7-17
Location in New Mexico Scattered near n border.
Location in Arizona Plateaus of n quarter.

DESERT creosotebush, mesquite, tarbush, catclaw acacia, paloverdes, bur-sages, cacti, desert saltbush
Elevation (feet) 100-4,500
Annual rainfall (inches) 3-15
Location in New Mexico Plains and valleys of s third.
Location in Arizona sw half and bottom of Grand Canyon.

Conspectus – Plant Families & Genera

The *Conspectus,* with each plant family and genus listed in the same order as the text, provides a quick means for identifying unknown trees. Listed first are the **Conifers**, followed by **Broadleaf Trees**, **Agaves and Cacti**, and finally, **Palms**. The number in the green square refers to the page number for either the tree genus or species.

CONIFERS

17
CUPRESSACEAE *(CYPRESS FAMILY)*
CYPRESS (CALLITROPSIS)

21
CUPRESSACEAE *(CYPRESS FAMILY)*
JUNIPER (JUNIPERUS)

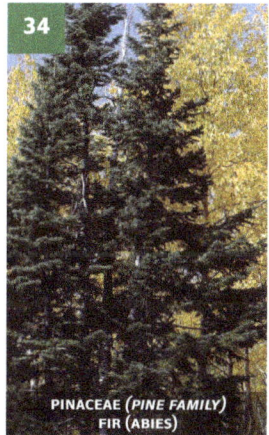

34
PINACEAE *(PINE FAMILY)*
FIR (ABIES)

38
PINACEAE *(PINE FAMILY)*
SPRUCE (PICEA)

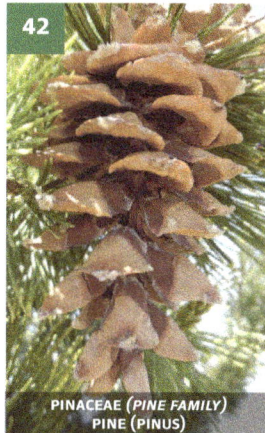

42
PINACEAE *(PINE FAMILY)*
PINE (PINUS)

62
PINACEAE *(PINE FAMILY)*
DOUGLAS-FIR (PSEUDOTSUGA)

BROADLEAF TREES

64

ADOXACEAE (*MUSKROOT FAMILY*)
MEXICAN ELDER (*SAMBUCUS*)

65

ANACARDIACEAE (*CASHEW FAMILY*)
SUMAC (*RHUS*)

68

BETULACEAE (*BIRCH FAMILY*)
ALDER (*ALNUS*)

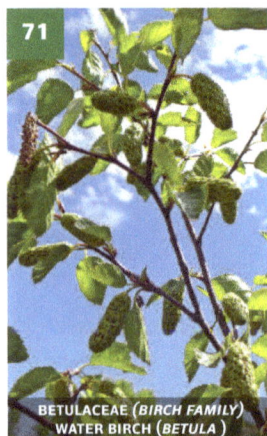

71

BETULACEAE (*BIRCH FAMILY*)
WATER BIRCH (*BETULA*)

72

BETULACEAE (*BIRCH FAMILY*)
HOP-HORNBEAM (*OSTRYA*)

73

BIGNONIACEAE (*BIGNONIA FAMILY*)
DESERT-WILLOW (*CHILOPSIS*)

75

BURSERACEAE (*BURSERA FAMILY*)
BURSERA (*BURSERA*)

78

CANNABACEAE (*HEMP FAMILY*)
NETLEAF HACKBERRY (*CELTIS*)

80

CELASTRACEAE (*BITTERSWEET FAMILY*)
CANOTIA (*CANOTIA HOLACANTHA*)

BROADLEAF TREES

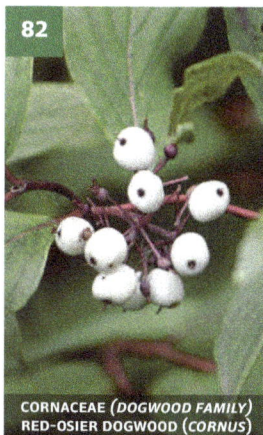

CORNACEAE *(DOGWOOD FAMILY)*
RED-OSIER DOGWOOD *(CORNUS)*

ERICACEAE *(HEATH FAMILY)*
MADRONE *(ARBUTUS)*

EUPHORBIACEAE *(SPURGE FAMILY)*
CASTOR-BEAN *(RICINUS)*

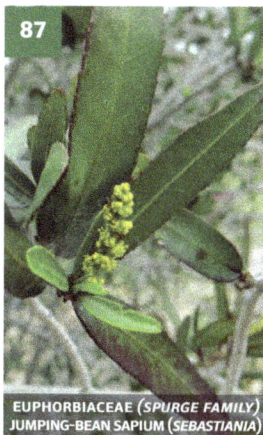

EUPHORBIACEAE *(SPURGE FAMILY)*
JUMPING-BEAN SAPIUM *(SEBASTIANIA)*

FABACEAE *(PEA FAMILY)*
CALIFORNIA REDBUD *(CERCIS)*

FABACEAE *(PEA FAMILY)*
MESCALBEAN *(DERMATOPHYLLUM)*

FABACEAE *(PEA FAMILY)*
CORALBEAN *(ERYTHRINA)*

FABACEAE *(PEA FAMILY)*
BIRD-OF-PARADISE *(ERYTHROSTEMON)*

FABACEAE *(PEA FAMILY)*
KIDNEYWOOD *(EYSENHARDTIA)*

BROADLEAF TREES

96

FABACEAE *(PEA FAMILY)*
HONEY-LOCUST *(GLEDITSIA)*

97

FABACEAE *(PEA FAMILY)*
LITTLELEAF LYSILOMA *(LYSILOMA)*

98

FABACEAE *(PEA FAMILY)*
TESOTA *(OLNEYA TESOTA)*

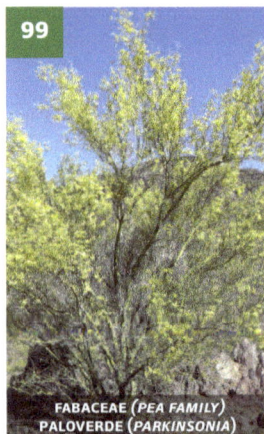

99

FABACEAE *(PEA FAMILY)*
PALOVERDE *(PARKINSONIA)*

104

FABACEAE *(PEA FAMILY)*
MESQUITE *(PROSOPIS)*

110

FABACEAE *(PEA FAMILY)*
SMOKETHORN *(PSOROTHAMNUS)*

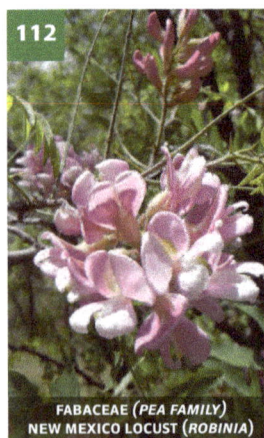

112

FABACEAE *(PEA FAMILY)*
NEW MEXICO LOCUST *(ROBINIA)*

114

FABACEAE *(PEA FAMILY)*
CATCLAW ACACIA *(SENEGALIA)*

116

FABACEAE *(PEA FAMILY)*
SWEET ACACIA *(VACHELLIA)*

BROADLEAF TREES

118

FAGACEAE *(BEECH FAMILY)*
OAK *(QUERCUS)*

144

JUGLANDACEAE *(WALNUT FAMILY)*
WALNUT *(JUGLANS)*

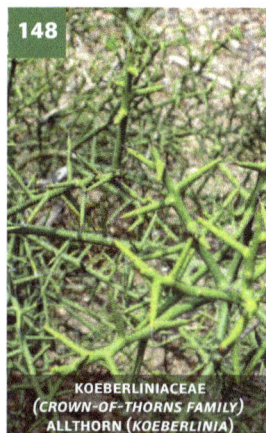

148

KOEBERLINIACEAE
(CROWN-OF-THORNS FAMILY)
ALLTHORN *(KOEBERLINIA)*

150

MALVACEAE *(MALLOW FAMILY)*
FREMONTIA *(FREMONTODENDRON)*

152

MORACEAE *(MULBERRY FAMILY)*
TEXAS MULBERRY *(MORUS)*

153

OLEACEAE *(OLIVE FAMILY)*
FORESTIERA *(FORESTIERA)*

154

OLEACEAE *(OLIVE FAMILY)*
ASH *(FRAXINUS)*

164

PLATANACEAE *(SYCAMORE FAMILY)*
ARIZONA SYCAMORE *(PLATANUS)*

166

RHAMNACEAE *(BUCKTHORN FAMILY)*
BITTER CONDALIA *(CONDALIA)*

BROADLEAF TREES

168

RHAMNACEAE *(BUCKTHORN FAMILY)*
BUCKTHORN *(RHAMNUS)*

174

RHAMNACEAE *(BUCKTHORN FAMILY)*
LOTEBUSH *(ZIZIPHUS OBTUSIFOLIA)*

176

ROSACEAE *(ROSE FAMILY)*
SERVICEBERRY *(AMELANCHIER)*

178

ROSACEAE *(ROSE FAMILY)*
MTN-MAHOGANY *(CERCOCARPUS)*

181

ROSACEAE *(ROSE FAMILY)*
HAWTHORN *(CRATAEGUS)*

183

ROSACEAE *(ROSE FAMILY)*
CHERRY *(PRUNUS)*

188

ROSACEAE *(ROSE FAMILY)*
QUININE-BUSH *(PURSHIA)*

190

ROSACEAE *(ROSE FAMILY)*
VAUQUELINIA *(VAUQUELINIA)*

192

RUSCACEAE
(BUTCHER'S-BROOM FAMILY)
BIGELOW NOLINA *(NOLINA)*

BROADLEAF TREES

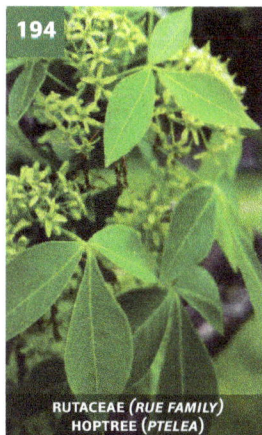

RUTACEAE *(RUE FAMILY)*
HOPTREE *(PTELEA)*

SALICACEAE *(WILLOW FAMILY)*
COTTONWOOD *(POPULUS)*

SALICACEAE *(WILLOW FAMILY)*
WILLOW *(SALIX)*

SAPINDACEAE *(SOAPBERRY FAMILY)*
MAPLE *(ACER)*

SAPINDACEAE *(SOAPBERRY FAMILY)*
WESTERN SOAPBERRY *(SAPINDUS)*

SAPOTACEAE *(SAPOTE FAMILY)*
GUM BUMELIA *(SIDEROXYLON)*

SIMAROUBACEAE *(AILANTHUS FAMILY)*
HOLACANTHA *(CASTELA EMORYI)*

SOLANACEAE *(NIGHTSHADE FAMILY)*
TREE TOBACCO *(NICOTIANA)*

TAMARICACEAE *(TAMARISK FAMILY)*
FRENCH TAMARISK *(TAMARIX)*

AGAVES & CACTI

231

ASPARAGACEAE *(ASPARAGUS FAMILY)*
YUCCA *(YUCCA)*

240

CACTACEAE *(CACTUS FAMILY)*
SAGUARO *(CARNEGIEA GIGANTEA)*

243

CACTACEAE *(CACTUS FAMILY)*
CHOLLA *(CYLINDROPUNTIA)*

250

CACTACEAE *(CACTUS FAMILY)*
SENITA *(PACHYCEREUS SCHOTTII)*

251

CACTACEAE *(CACTUS FAMILY)*
ORGANPIPE CACTUS *(STENOCEREUS)*

PALMS

253

ARECACEAE *(PALM FAMILY)*
CAL WASHINGTONIA *(WASHINGTONIA)*

CONIFERS

Conifers of the Southwest are *gymnosperms* (from the Greek meaning "naked seed"). Rather than being enclosed within an ovary, as in the angiosperms (e.g., 'wildflowers' and grasses), the seeds develop either on the surface of scales or leaves, which are often modified into cones. Cones are typically spherical to cylindric, composed of few to many woody scales, but in junipers the cones are fleshy and berrylike. Conifers of our region belong to either the Cypress Family (Cupressaceae: cypresses and junipers) or Pine Family (Pinaceae: pines, spruce, fir, and Douglas-fir); these may be separated in the key below. Foliage of our species is evergreen (persisting for more than one year) and either scalelike (Cypress Family) or needlelike (Pine Family).

1 Leaves needle-shaped, in fascicles of 1-5, each fascicle enclosed at base by a sheath . . .
. . . . **PINE** (*Pinus*)
1 Leaves single, linear or scalelike 2

2 Leaves mostly linear; leaves and fruit scales spirally arranged 3
2 Leaves mostly scalelike; leaves and fruit scales opposite 5

3 Base of leaves persistent on twigs as peglike projections (sterigmata) . **SPRUCE** (*Picea*)
3 Leaves not leaving peglike bases on twig upon falling 4

4 Leaves sessile; cones erect; buds rounded, resinous . . . . **FIR** (*Abies*)
4 Leaves narrowed at base into stalk; cones pendent with distinctive exserted bracts; buds pointed, non-resinous . . . . **DOUGLAS-FIR** (*Pseudotsuga*)

5 Cone berry-like; twigs not flattened; leaves scalelike or awl-shaped (both types usually present); seed wingless . . . . **JUNIPER** (*Juniperus*)
5 Cone woody; twigs flattened; leaves scalelike; seed winged . . . . . **CYPRESS** (*Callitropsis*)

CUPRESSACEAE *Cypress Family*

Both cypresses (*Callitropsis*) and junipers (*Juniperus*) are readily recognized by their minute, scalelike, appressed leaves, which completely clothe and obscure the small branchlets, and they are distinguished with ease by their cones. The woody cypress cone resembles a small walnut, and contains many seeds; the juniper cone is berrylike, usually fleshy, and usually with 1-3 seeds.

Callitropsis CYPRESS

Trees; leaves all alike, small, scalelike, closely imbricate (overlapping) and appressed to the branchlets, usually with a pit on the back containing a resin gland; cones nearly globular, with woody scales that separate at maturity, persistent on the branches several years; seeds winged, numerous under each scale.

1 Outer bark persistent except in saplings, rough . . . . **ARIZONA CYPRESS**
. . . . (*Callitropsis arizonica*)
1 Outer bark deciduous (except on the trunks of very old trees), leaving exposed the smooth, dark purplish-red inner bark. . . . . **SMOOTH ARIZONA CYPRESS**
. . . . (*Callitropsis glabra*)

Arizona Cypress *Callitropsis arizonica* (Greene) D.P. Little

ALSO CALLED rough-bark Arizona cypress

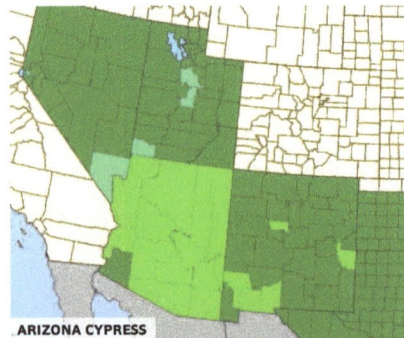

SYNONYMS *Cupressus arizonica* Greene, *Hesperocyparis arizonica* (Greene) Bartel

DESCRIPTION Scale-leaved evergreen tree usually about 30 feet tall, with straight trunk 1½ feet in diameter, up to 75 feet in height and 3 feet in diameter (maximum about 90 feet tall and 5½ feet in diameter). Crown conical or rounded.

TWIGS numerous, stout, 4-angled, branching nearly at right angles in all directions.

LEAVES scalelike, ¹⁄₁₆ inch (1.5 mm) long, pale blue green.

CONES short-stalked, ¾ to 1 inch (2 to 2.5 cm) in diameter, hard and woody, gray, with flattened scales bearing a point in center, remaining attached several years.

BARK variable, on small trunks smoothish and shedding in thin scales to expose dark red inner bark; or bark on larger trunks becoming rough and thick, furrowed and fibrous, or checkered, gray or blackish.

WOOD moderately soft and lightweight, light brown; formerly used for fence posts.

HABITAT Very scattered and local, forming groves in canyons and mountains of oak woodland, with evergreen oaks or junipers, 3,500 to 7,200 feet elevation.

NOTE Fine specimens of Arizona cypress can be seen in the Chiricahua National Monument in southeastern Arizona.

ARIZONA CYPRESS

ARIZONA CYPRESS

ARIZONA CYPRESS – CONES

ARIZONA CYPRESS

ARIZONA CYPRESS – LEAVES

Smooth Arizona Cypress *Callitropsis glabra* (Sudw.) Carrière

SYNONYMS *Cupressus glabra* Sudworth, *Hesperocyparis glabra* (Sudworth) Bartel

Smooth Arizona cypress is typically distinguished from *C. arizonica* by the smoothish outer bark shedding or peeling and exposing the dark red inner bark (though the largest trees may have thick, furrowed, gray bark). However the bark differences are not always clear-cut, and this 'species' may be better treated as *Callitropsis arizonica* var. *glabra*, as both forms are highly variable and show intergradation in shape of the crown (narrowly conic to broad and rounded), color of the foliage, size of the seeds, etc. Habitats are similar to that of Arizona Cypress. Smooth Arizona Cypress is widely cultivated as a symmetrical ornamental evergreen with bluish-green foliage. This cypress is also sometimes called "yew-wood," because its smooth purple-red bark resembles that of the northwestern yew, *Taxus brevifolia*.

The first reference to this cypress was published in 1895 and was based on the discovery of a grove in central Arizona by Prof. J. W. Tourney, who believed the tree to be a form of Arizona cypress. It was not separated from Arizona cypress until 1910, when it was named and described from a grove of trees discovered by Mr. Arthur H. Zachau in the Verde River Canyon, about 16 miles south east of the town of Camp Verde, Ariz. At that time, the grove covered an area about 6 miles long by 1½ miles wide.

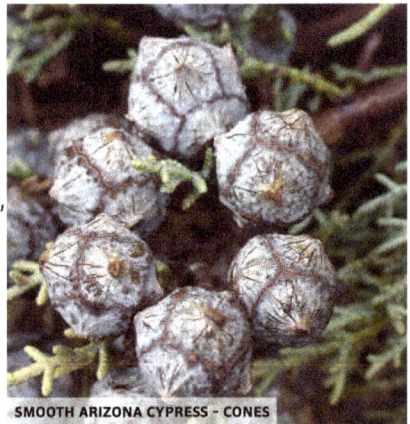

SMOOTH ARIZONA CYPRESS - CONES

SMOOTH ARIZONA CYPRESS - LEAVES

SMOOTH ARIZONA CYPRESS

Juniperus JUNIPER

Evergreen trees or shrubs; leaves small, in alternate pairs or whorls, subulate and spreading in the juvenile form, imbricate and appressed in the mature form (except in the low-growing shrub *Juniperus communis*); flowers commonly monoecious; cones berry-like, the scales becoming fleshy and not separating at maturity, or only slightly so at apex of the cone.

Arizona and New Mexico have extensive stands of juniper, chiefly on well-drained, rather sterile soils. The wood was formerly much used in manufacturing lead pencils, but today its chief use is for firewood and for fence posts, for which the very durable heartwood is well-suited (Utah juniper [*J. osteosperma*], with relatively straight branches, is especially suitable for fence posts). The foliage of junipers is browsed when other forage is scarce but is injurious to livestock if eaten too freely. The berries are eaten by birds and other wild creatures, and formerly were used as food by the region's Native Americans. The fruit of *J. communis* (a low-growing shrub not treated here) are used elsewhere to give the characteristic flavor to gin, and are the source of *oil of juniper,* which has been used extensively in patent medicines. Juniper is used in various ways for medicinal and ritualistic purposes by the Hopi. *Juniperus* is the Latin name for Juniper.

1 Trunk bark in thick, squarish plates; fruit red-brown, ½ inch (12 mm) diameter, usually 4-seeded, ripening in 2 years............ ALLIGATOR JUNIPER (*Juniperus deppeana*)
1 Trunk bark fibrous and shreddy 2

2 Fruit bright red to red-brown beneath the whitish bloom 3
2 Fruit bluish to blue-black, 1-3 seeded 7

3 Fruit bright red (rarely copper-colored) ¼ inch (6 mm) diameter, 1-seeded 4
3 Fruit dull red-brown or copper-colored 5

4 Fruit bright red or orange; seed with large, dark, ridged band and 3 concavities ....
.... REDBERRY JUNIPER (*Juniperus coahuilensis*)
4 Fruit copper-colored; seed without distinctive marking............ PINCHOT JUNIPER
.... (*Juniperus pinchotii*)

5 Fruit ¼ to ¾ inch (6 to 20 mm) diameter; seed marked at base by hilum; fruit maturing in 1 year............................. ONE-SEED JUNIPER (*Juniperus monosperma*)
5 Fruit ⅙ to ⅓ inch (4 to 8 mm) diameter; seed completely enclosed; fruit maturing in 2 years; heartwood brown.. 6

6 Leaves in whorls of 3's, rarely opposite; seed marked at base by light-colored hilum ...
.... CALIFORNIA JUNIPER (*Juniperus californica*)
6 Leaves opposite, rarely in 3's; seed marked to middle by conspicuous hilum...........
.... UTAH JUNIPER (*Juniperus osteosperma*)

7 Leaf margin smooth; heartwood distinctly reddish ...... ROCKY MOUNTAIN JUNIPER
.... (*Juniperus scopulorum*)
7 Leaf margin minutely fringe-toothed (under hand lens); heartwood brownish ....
.... ONE-SEED JUNIPER (*Juniperus monosperma*)

California Juniper *Juniperus californica* Carr.

SYNONYM *Juniperus cedrosiana* Kellogg

DESCRIPTION Usually a shrub but occasionally a small tree to 25 feet tall; trunk usually with multiple stems, the crown rounded.

TWIGS round.

LEAVES scalelike, ⅟₃₂ to ⅛ inch (1–3 mm) long, light green with a conspicuous gland, not or only slightly overlapping, closely appressed to the twigs.

CONE round, ⁵⁄₁₆ to ½ inch (8–12 mm) diameter, glaucous bluish brown, mostly 1-seeded; maturing in 1 year.

BARK smooth, brown to gray when young, becoming gray and shedding in thin strips.

CALIFORNIA JUNIPER

HABITAT Dry, rocky slopes and flats, from low-elevations to about 5,200 feet.

NOTE Often found with singleleaf Pinyon in woodlands and with Joshua-tree where drier; more common in California, entering our range in northwestern arizona.

Redberry Juniper *Juniperus coahuilensis* (Martínez) Gaussen ex R.P. Adams

SYNONYM *Juniperus arizonica* (R.P. Adams) R.P. Adams

DESCRIPTION Shrub or small tree to 25 feet high, branched at base or with a single stem to 3 feet diameter; crown round to irregular; able to resprout after fire or cutting.

TWIGS 3- or 4-sided in cross-section.

LEAVES green to light green, ⅟₃₂ to ⅛ inch (1 to 3 mm) long, not overlapping or overlapping by up to ¼ of their length, with conspicuous glands, at least ¼ of which have a white crystalline exudate.

CONE round to ovoid, ¼ to ⁵⁄₁₆ inch (6–7 mm) wide, glaucous, yellow-orange to dark red, fleshy and somewhat sweet, usually 1-seeded; maturing in 1 year.

BARK brown to gray, at first smooth, on mature trunk and large branches shedding in long, ragged strips (or sometimes in flakes).

CALIFORNIA JUNIPER

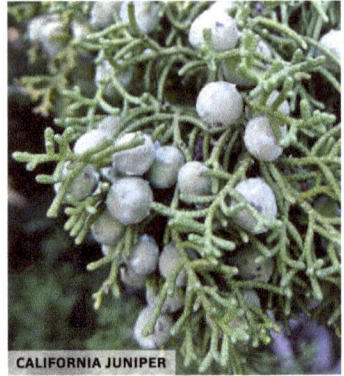

CALIFORNIA JUNIPER

WOOD soft, lightweight, light red with whitish sapwood; used for fence posts and fuel.
HABITAT Most often in grasslands, where it may be the only tree species present; sometimes on rocky slopes; to about 5,000 feet elevation.
NOTE *Juniperus coahuilensis* is very similar to **one-seed juniper** (*Juniperus monosperma*). The most notable difference is that the cones ("berries") of redberry juniper are pinkish, while those of *Juniperus monosperma* are purplish; there is also more resin on the leaves of *Juniperus monosperma.* In Texas, redberry juniper intergrades with Pinchot juniper (*Juniperus pinchottii*) where they overlap.

REDBERRY JUNIPER

REDBERRY JUNIPER

REDBERRY JUNIPER

REDBERRY JUNIPER

Alligator Juniper *Juniperus deppeana* Steud.

ALSO CALLED western juniper (lumber)

SYNONYM *Juniperus pachyphloea* Torr.

DESCRIPTION Medium-sized, scale-leaved evergreen tree usually 20 to 40 feet tall with a large, short trunk 2 to 3 feet in diameter, reaching a maximum height of 65 feet and a trunk diameter of 7 feet. Crown rounded and spreading, or in age irregular and partly dead.

LEAFY TWIGS 1/32 to 1/16 inch (1 to 1.5 mm) in diameter.

OLDER TWIGS reddish brown and nearly smooth, peeling off.

LEAVES scalelike, 1/16 long (1.5 mm), blue green, glandular, mostly with a whitish resin drop or gland, or on leading twigs needlelike, up to 1/4 inch (6 mm) long, pale or whitish.

CONE berry-like, 1/2 inch (12 mm) in diameter, bluish or brownish, covered with a bloom, hard and mealy, 3- or 4-seeded, maturing the second year.

BARK thick and rough, deeply furrowed into checkered or square plates, gray or blackish, suggesting the back of an alligator.

WOOD soft, lightweight, light red with narrow whitish sapwood; used for fuel and fence posts.

HABITAT Common but usually scattered on hillsides and mountains in oak woodland, pinyon-juniper woodland, and lower part of ponderosa pine forest, 4,500 to 8,000 feet elevation.

NOTE Alligator juniper, the largest juniper in New Mexico and Arizona, is usually scattered rather than in pure stands. The trees attain a great age

ALLIGATOR JUNIPER - BARK

ALLIGATOR JUNIPER

and develop very large trunks. With some weathered dead limbs attached and with dead strips vertically ascending the trunk and branches, older trees have a grotesque appearance. Sprouts often form at the base of stumps. Juniper seeds are widely spread by birds and wild mammals which eat the "berries."

ALLIGATOR JUNIPER

ALLIGATOR JUNIPER

ALLIGATOR JUNIPER – LEAVES

One-Seed Juniper *Juniperus monosperma* (Engelm.) Sarg.

SYNONYM *Juniperus occidentalis* var. *gymnocarpa* Lemmon

DESCRIPTION Much-branched, spreading, often scraggy, scale-leaved evergreen shrub or small tree 10 to 25 feet tall, with several curved branches from the ground, usually without a single upright trunk but sometimes with a trunk to 1½ feet in diameter. Pollen and seeds borne on different trees (dioecious).

LEAFY TWIGS stout, about ¹⁄₁₆ inch (1.5 mm) or less in diameter.

LEAVES scalelike, ¹⁄₁₆ inch (1.5 mm) or more in length, yellow green.

CONE berry-like, ¼ inch (6 mm) in diameter, dark blue, covered with a bloom, juicy, 1-seeded, maturing in one year.

BARK fibrous and shreddy, gray.

WOOD soft, lightweight, light reddish brown with whitish sapwood; much used for fence posts and fuel.

HABITAT Common and widespread on plains, plateaus, and foothills in pinyon-juniper woodland, growing with pinyon and Utah juniper, or sometimes in upper part of desert and desert grassland, 3,000 to 7,000 feet elevation, widely distributed.

NOTE One-seed juniper is the commonest juniper in New Mexico but not as abundant in Arizona as Utah juniper. These two similar species occasionally grow together but can usually be distinguished by the following differences: Utah juniper

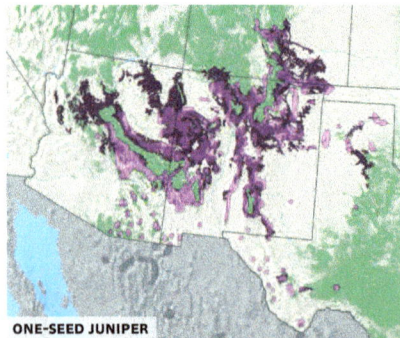

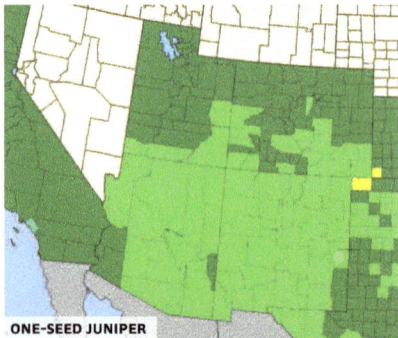

ONE-SEED JUNIPER

ONE-SEED JUNIPER

is larger, with a definite trunk, while one-seed juniper is smaller and usually shrubby, with several branches from the ground. Utah juniper has larger, mealy, 1- or 2-seeded "berries" borne on the same trees as the pollen (monoecious), while one-seed juniper has smaller, juicy, 1-seeded "berries" on the female trees and pollen on male trees (dioecious).

ONE-SEED JUNIPER

ONE-SEED JUNIPER

Utah Juniper *Juniperus osteosperma* (Torr.) Little

ALSO CALLED western juniper (lumber)

SYNONYM *Juniperus utahensis* (Engelm.) Lemmon

DESCRIPTION Small, scale-leaved evergreen tree 15 to 40 feet tall, usually with definite upright trunk 1 to 3 feet or more in diameter, branching usually several feet above the ground to form a broad, rounded or conical, open crown.

LEAFY TWIGS stout, about ¹⁄₁₆ inch (1.5 mm) or less in diameter.

LEAVES scalelike, about ¹⁄₁₆ inch (1.5 mm) long, yellowgreen.

CONE berry-like, ¼ to ⅝ inch (6 to 15 mm) in diameter, brownish, covered with a bloom, mealy, 1- or 2-seeded.

BARK fibrous and shreddy in long strips, gray.

WOOD soft, lightweight, light brown with thick whitish sapwood; wood durable for fence posts, also used for fuel.

HABITAT Common to abundant on dry plains, plateaus, hills, and mountains in pinyon-juniper woodland, often in pure stands or with pinyon, 3,000 to 7,500 feet elevation.

NOTE Utah juniper is the commonest juniper in Arizona but in New Mexico is of restricted occurrence. In various places it occupies pure stands, especially westward

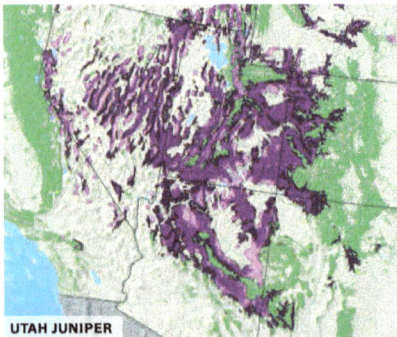

UTAH JUNIPER

UTAH JUNIPER

and at elevations below the limits of pinyon. Within the past 100 years junipers have increased, thickened, and spread onto adjacent short grass and desert grassland vegetation in the Southwest. This change is considered undesirable on lands used for grazing. The large, mealy "berries" are eaten by wildlife.

TIP The thicker twigs and larger and more glaucous cones ("berries") help distinguish Utah juniper from Rocky Mountain juniper.

UTAH JUNIPER

UTAH JUNIPER (FRONT), SINGLELEAF PINYON PINE (REAR)

Pinchot Juniper *Juniperus pinchotii* Sudworth

SYNONYM *Juniperus erythrocarpa* Cory, *Juniperus monosperma* var. *pinchotii* (Sudworth) Van Melle

DESCRIPTION Scale-leaved evergreen shrub or small tree to about 20 feet tall; trunk usually with multiple stems, each to about 8 inches diameter; crown irregular. **TWIGS** stiff, 3- or 4-sided in cross-section, about ¹⁄₃₂ inch (1 mm) diameter. **LEAVES** scalelike (needlelike on leading twigs), about ¹⁄₁₆ inch long (1.5 mm), yellow-green, not overlapping (or overlapping only slightly), many leaves with a fragrant white exudate.

CONE berry-like, ¼ to ⅜ inch (6 to 9 mm) in diameter, coppery-brown, juicy, sweet, and not resinous, maturing the first year; mostly 1-seeded.

BARK at first smooth, becoming flaky; on older trees fibrous and pale gray, exfoliating in strips. **WOOD** soft, lightweight, light red with whitish sapwood; used for fence posts and fuel.

HABITAT Usually on gravelly, calcareous soils, often with honey mesquite (*Prosopis glandulosa*) and shrubby oaks (*Quercus*); from low-elevations to about 3,500 feet. Also in northern Mexico.

NOTE Stumps can resprout following fire or cutting.

PINCHOT JUNIPER

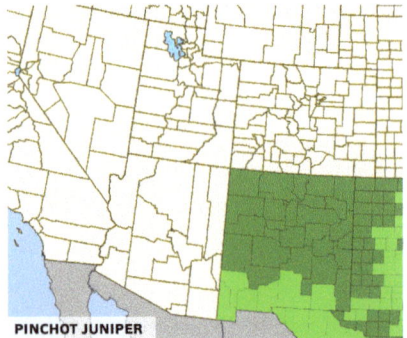

PINCHOT JUNIPER

PINCHOT JUNIPER

ETYMOLOGY named in honor of *Gifford Pinchot* (1865-1946), first chief of the United States Forest Service. It is a common tree at Palo Duro State Park near Canyon, Texas, where it was discovered.

PINCHOT JUNIPER - MALE CONES

PINCHOT JUNIPER

Rocky Mountain Juniper *Juniperus scopulorum* Sarg.

ALSO CALLED Rocky Mountain redcedar, western juniper (lumber)

DESCRIPTION Small to medium-sized, scale-leaved evergreen tree 20 to 50 feet tall, with straight trunk up to 1½ feet in diameter, with narrow and pointed, open, conical crown, and with slender branches often drooping at the ends.

LEAFY TWIGS slender, about 1/32 inch (1 mm) in diameter.

LEAVES scalelike, ¹⁄₁₆ inch (1.5 mm) long, usually gray green, or on leading shoots needlelike, up to ¼ inch (6 mm) long.

CONE berry-like, ¼ inch (6 mm) in diameter, bright blue, covered with a bloom, juicy, usually 2-seeded, maturing the second year.

BARK thin, fibrous and shreddy, dark reddish brown or gray.

WOOD soft, lightweight, deep red with thick whitish sapwood. The wood is used for fence posts, fuel, and lumber, and is suitable for cedar chests.

HABITAT Scattered in mountains and canyons of pinyon-juniper woodland and lower part of ponderosa pine forest, 5,000 to 9,000 feet elevation.

NOTE Rocky Mountain juniper, a slender tree with grayish green foliage, is graceful and highly ornamental. The form with drooping twigs is called "weeping juniper." The trees grow faster than other southwestern junipers and are planted in shelter belts and as ornamentals.

ROCKY MOUNTAIN JUNIPER

ROCKY MOUNTAIN JUNIPER

ROCKY MOUNTAIN JUNIPER

ROCKY MOUNTAIN JUNIPER

ROCKY MOUNTAIN JUNIPER

PINACEAE *Pine Family*

Large evergreen trees with needle-shaped leaves; male flowers in short catkins; female flowers in scaly catkins, these becoming cones, with 2 or more ovules at the base of each scale; fertile scales numerous, spirally imbricated. See key to Conifers, page 17.

Abies FIR

Large trees with spreading or ascending branches; leaves flat, blunt, short, so arranged as to make the branches appear flat; cones erect, cylindrical, borne near the top of the tree, their scales thin and deciduous. *Pinus* is the classical Latin name.

1   Needles 2 to 3 inches (5 to 7.5 cm) long; cones gray-green, scales slightly wider than long . . . . **WHITE FIR** (*Abies concolor*)
1   Needles on lower branches ¾ to 1¾ inches (2 to 4 cm) long; cones dark purple, scales slightly longer than broad 2

2   Mature bark hard, tough, smooth, gray . . . . **SUBALPINE FIR** (*Abies lasiocarpa*)
2   Mature bark distinctly soft and corky, yellow-white . . . . **CORKBARK FIR** . . . . (*Abies lasiocarpa* var. *arizonica*)

White Fir   *Abies concolor* (Gord. & Glend.) Lindl. ex Hildebr.

ALSO CALLED **balsam fir, silver fir, white balsam**
DESCRIPTION Large, needle-leaved evergreen tree to 150 feet in height and 3½ feet in trunk diameter, with pointed conical crown becoming irregular in age.
NEEDLES spreading and curved upward, flat, 1½ to 2½ inches (3.5 to 6 cm) long, usually blunt, pale blue green or silvery.
CONES in top of tree, upright, 3 to 5 inches (7.5 to 12.5 cm) long, usually grayish green, with scales falling apart at maturity.
BARK on small trunks smoothish, gray, becoming very thick, hard, and deeply furrowed into scaly ridges.
WOOD soft, very lightweight, whitish or light brown.

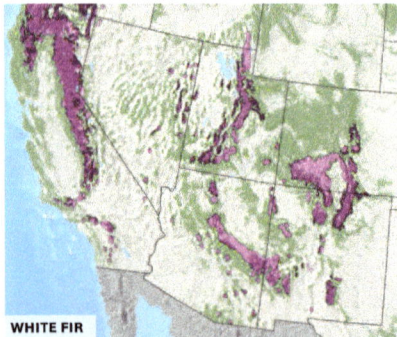

WHITE FIR

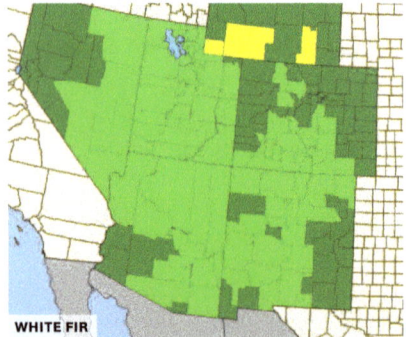

WHITE FIR

HABITAT Common in ponderosa pine, Douglas-fir, and spruce-fir forests, 5,500 to 10,000 feet, in high mountains.

NOTE White fir is associated with other coniferous species in forests at high elevations. Limited quantities of white fir are cut for lumber in the Southwest, often sold with lumber of other species. The trees are suitable as ornamentals and shade trees.

TIP *Abies concolor* is distinguished from *A. lasiocarpa* by its dark gray and furrowed bark on mature trees, longer needles on the lower branches, and yellowish green to greenish purple or grayish green female (seed) cones.

WHITE FIR

WHITE FIR

WHITE FIR - CONES

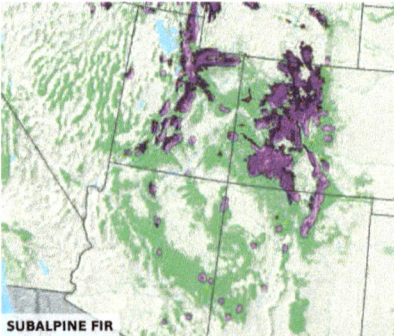

Subalpine Fir　*Abies lasiocarpa* (Hook.) Nutt.

ALSO CALLED alpine fir, white balsam, white fir (lumber)

SYNONYM *Abies arizonica* Merriam, *Abies bifolia* A. Murr.

DESCRIPTION Large, needle-leaved evergreen tree to 90 feet tall and 3 feet in trunk diameter, or larger; but dwarfed and shrubby at timber line. Crown long, narrow, and sharp-pointed, with branches extending nearly to base of tree.

NEEDLES spreading or sometimes in 2 rows, flat, 1 to 1¾ inches (2.5 to 4 cm) long, blunt, or on uppermost branches pointed and shorter, dark blue green.

CONES in top of tree, upright, 2½ to 4 inches (6 to 10 cm) long, dark purple, finely hairy, with scales falling apart at maturity.

BARK becoming fissured and scaly, gray; in the variety corkbark fir, bark soft, spongy or corky, smoothish, thin, creamy white.

SUBALPINE FIR

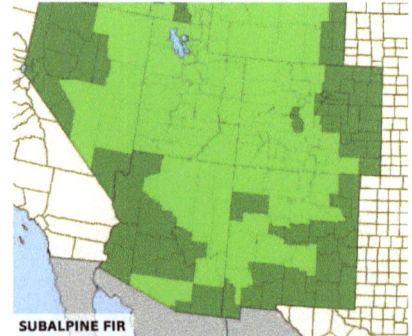

WOOD soft, lightweight, light brown.

HABITAT Characteristic, usually common tree of spruce-fir forest up to timberline, growing with Engelmann spruce, 8,000 to 12,000 feet elevation, in higher mountains.

NOTE In the Southwest, subalpine fir, including corkbark fir, is too restricted to be of much importance for lumber, though some is cut in the White Mountains of eastern Arizona and elsewhere. This species is popular as an ornamental and as a Christmas tree.

ADDITIONAL SPECIES Corkbark fir (*Abies lasiocarpa* var. *arizonica* (Merriam) Lemmon; also called Arizona fir.

Readily distinguished by its peculiar, whitish, corky bark and very glaucous foliage, occurs on San Francisco Mountain, Arizona, where it was first discovered, and elsewhere on scattered mountains in Arizona, New Mexico, and southwestern Colorado. Because of its attractive, showy bark, it is highly valued as an ornamental tree.

CORKBARK FIR

SUBALPINE FIR

SUBALPINE FIR

SUBALPINE FIR

CORKBARK FIR – BARK

CORKBARK FIR

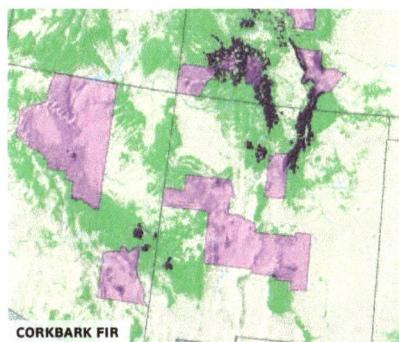
CORKBARK FIR

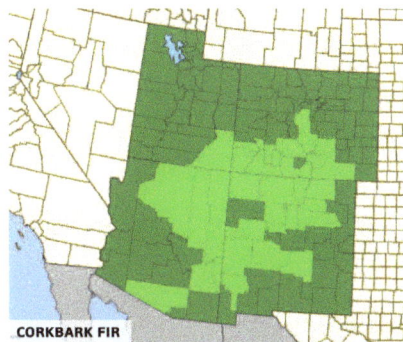
CORKBARK FIR

Picea SPRUCE

Conical trees with short stiff sharp-pointed solitary leaves standing out in all directions from the stems; cones pendulous, their scales rather thin, persistent, the bracts shorter than the scales. *Picea* is from the Greek *pix,* pitch; in allusion to the exuded resin of these trees.

1   Cone 1 to 2½ inches (2.5 to 6 cm) long, persistent 1 year; needles flexible, acute but not prickly to touch; twigs minutely hairy; buds ⅛ to ¼ inch (3 to 6 mm) long with scales usually appressed; bark on mature trees thin and scaly . . . . ENGELMANN SPRUCE (*Picea engelmannii*)

1   Cone 2¼ to 4½ inches (5.5 to 11 cm) (mostly about 3½ inches [8.5 cm]) long, persistent 2 years; needles stiff, bristle-pointed; twigs glabrous; buds ¼ to ½ inch (6 to 12 mm) long with scales usually reflexed; bark on mature trees thick and furrowed . . . . BLUE SPRUCE (*Picea pungens*)

Engelmann Spruce   *Picea engelmannii* Parry ex Engelm.

ALSO CALLED white spruce, mountain spruce, silver spruce

DESCRIPTION Large, needle-leaved evergreen tree to 80 feet or more in height, with straight trunk to 3 feet or more in diameter, and with narrow, pointed, conical crown and horizontal or slightly drooping branches extending nearly to ground; or at timber line dwarfed and bushy.

TWIGS roughened by peglike bases of fallen needles; twigs and leaf bases usually hairy.

NEEDLES 4-angled, ⅝ to 1¼ inches (15 to 30 mm) long, pointed but not stiff, dark or pale blue green, with disagreeable odor when crushed.

CONES 1½ to 2½ inches (3.5 to 6 cm) long, light brown, with papery scales more or less rounded and distinctly thinner at tip.

BARK thin, with loosely attached scales or flakes, grayish or purplish brown.

WOOD soft, lightweight, light yellow to reddish brown. The lumber is used for building construction and boxes, and the wood is also a source of mine timbers, railroad ties, and poles; desirable as an ornamental in cool moist climates.

ENGELMANN SPRUCE - BARK

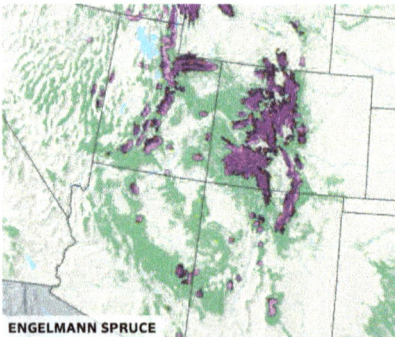
ENGELMANN SPRUCE

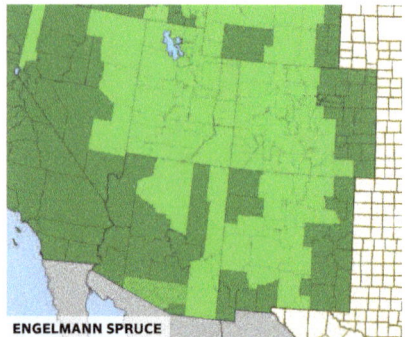
ENGELMANN SPRUCE

HABITAT A common and characteristic tree of the spruce-fir forest up to the timber line, often growing crowded in dense stands, pure or mixed with alpine fir, Douglas-fir, white fir, and blue spruce, usually between 9,000 feet and the timber line up to 12,000 feet elevation but occasionally as low as 8,000 feet, in higher mountains. Engelmann spruce is an important tree species at higher elevations in mountains of the Southwest, though limited in area and accessibility.

ETYMOLOGY This species honors *George Engelmann* (1809-84), German-American physician and botanist of St. Louis, an authority on conifers, yuccas, cacti, and other plant groups. He was first to call attention to the immunity of American grape stocks to phylloxera, which played a large role in saving the European wine industry.

ENGELMANN SPRUCE

ENGELMANN SPRUCE

ENGELMANN SPRUCE - NEEDLES

Blue Spruce   *Picea pungens* Engelm.

ALSO CALLED Colorado blue spruce, Colorado spruce, silver spruce

SYNONYM *Picea parryana* (Andre) Sarg.

DESCRIPTION Large, needle-leaved evergreen tree up to 80 feet tall and 2 feet in trunk diameter or larger, with conical crown of bluish foliage, at least on young trees and parts.

TWIGS roughened by peg-like bases of fallen needles; twigs and leaf bases usually not hairy.

NEEDLES 4-angled, ¾ to 1⅛ inches (2 to 3 cm) long, stiff and spine-pointed, dull blue green or silvery blue or becoming darker on older parts.

CONES 2½ to 4 inches (6 to 10 cm) long, light brown, with scales more or less straight across tip and not thinner.

BARK rough and thick, furrowed into scaly ridges, gray or brown.

WOOD soft, lightweight, brownish or whitish.

HABITAT Uncommon in spruce-fir forest and Douglas-fir forest, but sometimes in dense stands, 7,000 to 11,000 feet elevation, in higher mountains.

NOTE Blue spruce is less common than Engelmann spruce and less widely distributed, often occurring at slightly lower elevations. The lumber of both species is cut and marketed together. The trees are extensively planted for ornament because of the bluish, often silvery, foliage and also used in shelter belts. Blue spruce is the state tree of both Colorado and Utah.

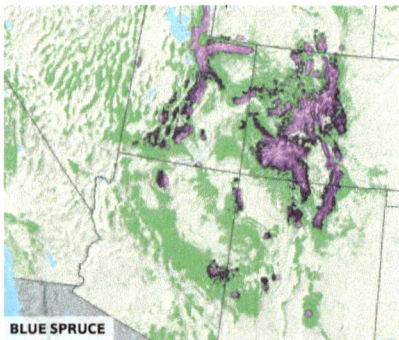

BLUE SPRUCE

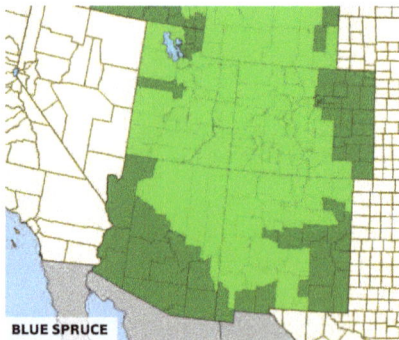

BLUE SPRUCE

BLUE SPRUCE

BLUE SPRUCE - CONE

BLUE SPRUCE

Pinus PINE

Large or small trees with needle-shaped leaves in fascicles of 2 or more, surrounded by a persistent or deciduous sheath at the base. Flowers are male and female, usually borne on different branches of the same tree. Male flowers, which produce pollen, are short, oval, and budlike, or long cylindrical bodies, clustered at the ends of mature leafy branches. Female flowers, which produce cones and seed, are small, greenish, scaly, and conelike, produced singly or in pairs (or larger groups) near the ends of young growing shoots. Fruits are woody, scaly cones, maturing in 2 or 3 years. Each of the scales in the central portion of the cone usually bears 2 seeds at its base. The cones of some pines remain on the trees only a few weeks after ripening, while those of others persist for many years. Within a few weeks after maturing, most pine cones open under the heat of the sun and free their seeds. The cones of a few pines, however, may remain closed for several or for many seasons, sometimes opening fully only under the heat of a forest fire. *Pinus* is the ancient Latin name.

1   Needles in clusters of 5 (except Pinyons), the cluster sheath deciduous and absent on mature needles; one fibrovascular bundle in cross section of needle; cone scales without prickles (except in bristlecone pine) (Soft Pines) 2

1   Needles in clusters of 2 or 3 (mostly 5 in Arizona Pine), the cluster sheath persistent (deciduous in Chihuahuan Pine); 2 fibro-vascular bundles in cross section of needle; cone scales usually armed with prickles (Hard Pines) 6

2   Needles in clusters of 1 to 4; cones globose, few-scaled; seed large and edible, without wing (Pinyon Pines) 3

2   Needles in clusters of 5; cones many-scaled 4

3   Needles single . . . . **SINGLELEAF PINYON** (*Pinus monophylla*)

3   Needles predominantly in clusters of 2 . . . . **TWO-NEEDLE PINYON** (*Pinus edulis*)

3   Needles in clusters of 3 . . . . **MEXICAN PINYON** (*Pinus cembroides*)

4   Cones scales with long, slender prickles; cones 2½ to 4 inches (6 to10 cm) long . . . .
    . . . . **BRISTLECONE PINE** (*Pinus aristata*)

4   Cone scales without prickles; cones 3-10 inches (7.5 to 25 cm) long 5

5   Leaves entire . . . . **LIMBER PINE** (*Pinus flexilis*)

5   Leaves serrulate . . . . **SOUTHWESTERN WHITE PINE** (*Pinus strobiformis*)

6   Needles in clusters of 5 . . . . **ARIZONA PINE** (*Pinus arizonica*)

6   Needles in clusters of 2 and 3 (or only 3) 7

7   Needles in clusters of 3, 2 to 4 inches (5 to 10 cm) long, cluster sheath deciduous; cone 1½ to 2 inches (3.5 to 5 cm) long, maturing in 3 years, long-stalked, often remaining closed, long persistent . . . . **CHIHUAHUAN PINE** (*Pinus leiophylla*)

7   Needles mostly in clusters of 3 (sometimes 2), more than 4 inches (10 cm) long, sheath persistent; cone 3 to 6 inches (7.5 to 15 cm) long, maturing in 2 years, short-stalked, opening, deciduous 8

8   Needles 4 to 11 inches (10 to 28 cm) (mostly 5 to 7 inches, (12.5 to 17.5 cm)) long, in clusters of 2 and 3, yellow-green, not drooping . **PONDEROSA PINE** (*Pinus ponderosa*)

8   Needles 8 to 15 inches (20 to 37.5 cm) long, in clusters of 3 (rarely 2 to 5), dark green, often drooping . . . . **APACHE PINE** (*Pinus engelmannii*)

Bristlecone Pine *Pinus aristata* Engelm.

ALSO CALLED foxtail pine

DESCRIPTION Small needle-leaf evergreen tree 40 feet or less in height and up to 2½ feet in trunk diameter, with broad, irregular crown and spreading branches, or a low bushy shrub at the timber line.

NEEDLES numerous and densely crowded, 5 in a bundle, 1 to 1½ inches (2.5 to 3.5 cm) long, stout, dark green, curved and pressed against the twig and not spreading out, remaining attached 10 to 15 years and forming brushlike or "foxtail" masses along the ends of the branches.

CONES 2½ to 4 inches (6 to 10 cm) long, dark purplish brown, each scale with a slender prickle nearly ¼ inch (6 mm) long.

BARK on small trunks and branches smooth and whitish, on larger trunks becoming irregularly fissured, scaly, and reddish brown.

WOOD soft, moderately heavy, brownish red; formerly used locally for fuel and mine props.

HABITAT Very local and widely scattered, in open grassy stands or in spruce-fir forest up to timber line, 9,700 to 11,500 feet elevation, on the highest mountains of northern New Mexico and Colorado.

NOTE Bristlecone pine, named from the prickles on the cones, has an unusual, scattered distribution on widely separated high peaks.

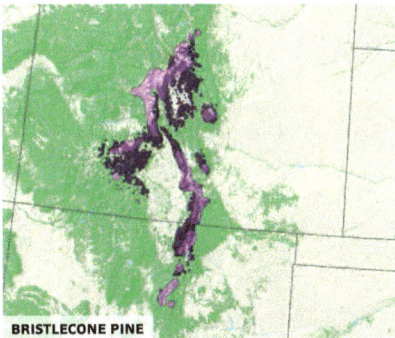

BRISTLECONE PINE

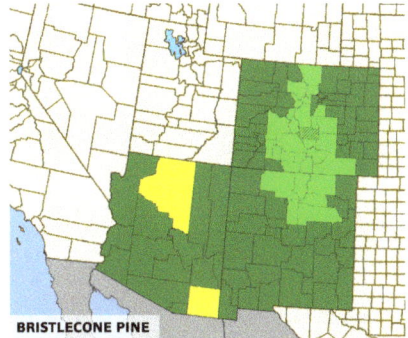

BRISTLECONE PINE

BRISTLECONE PINE

BRISTLECONE PINE

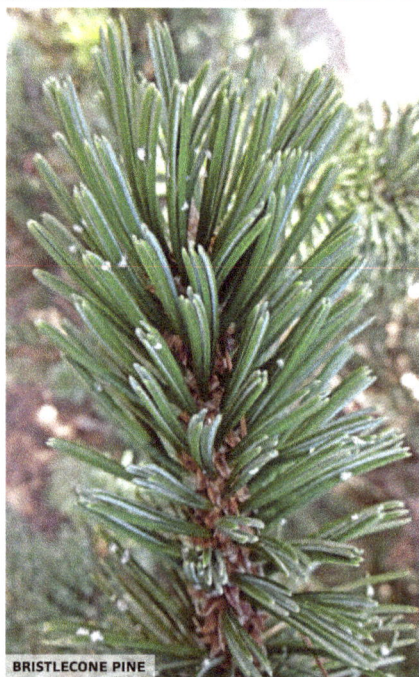

BRISTLECONE PINE

Arizona Pine *Pinus arizonica* Engelm.

ALSO CALLED **Arizona ponderosa pine, Arizona yellow pine**

SYNONYM *Pinus ponderosa* var. *arizonica* (Engelm.) Shaw

DESCRIPTION Large, evergreen tree 30-35 m tall, to 1 m in diameter; crown thick and rounded in mature trees. Branches thick and strong, lower ones drooping, upper ascending.

NEEDLES Borne in clusters of usually 5 needles (varies from 3 to 5), stiff, erect, 5 to 9 inches (12.5 to 22 cm) long, growing in groups at the end of branchlets. Stomata are present on the dorsal and ventral surfaces, margins finely serrate, with 6-10 resin canals; fascicle sheaths brown, up to ⅝ inch (15 mm) long.

CONES Ovoid to conical, symmetrical, erect to slightly reflexed, 2½ to 3½ inches (6-9 cm) long, borne singly, or in 2's or 3's on short, stout peduncles, reddish brown; scales stiff about ½ to ⅝ inch (12 to 15 mm) wide, apical margin rounded and smooth, transversely keeled, ashy gray and bearing a sharp persistent, recurved prickle.

BARK irregularly fissured with reddish, brown scaly plates.

WOOD hard, yellowish with whitish sapwood.

HABITAT Found on deep, well-drained soils in valleys, on mesas, and in the mountains from 6,500-9,500 feet.

NOTE Arizona Pine was previously considered a variety of ponderosa pine (*P. ponderosa* var. *arizonica*); distinguishing features in the field are: 3 to 5 (usually 5) needles in *P. arizonica* vs. only 2 to 3 in *P. ponderosa;* and cone scales in *P. arizonica* have a small recurved prickle while those of *P. ponderosa* have a large, strong, erect prickle.

ARIZONA PINE

ARIZONA PINE

Mexican Pinyon *Pinus cembroides* Zucc.

ALSO CALLED nut pine, pinyon pine, Mexican pinyon pine

SYNONYM *Pinus cembroides* var. *bicolor* Little

DESCRIPTION Small, needle-leaf evergreen tree 15 to 30 feet or more in height, with trunk 1.5 feet or more in diameter and with compact, rounded, spreading crown.

NEEDLES 3 in a bundle, 1 to 1¾ inches (2.5 to 4 cm) long, slender, dark blue green.

CONES egg-shaped, 1 to 1½ inches (2.5 to 3.5 cm) long, light brown, with thick blunt scales and hard-shelled, brown seeds ⅜ inch (9 mm) long, oily and edible and known as pinyon nuts or *piñones*.

BARK deeply furrowed into scaly ridges, reddish brown or blackish, on large trunks with flattened plates.

WOOD soft, heavy, resinous, yellow; the pitchy wood is used as firewood.

HABITAT Characteristic and common tree of the pinyon-juniper woodland, associated with junipers and oaks, 5,000 to 7,500 feet elevation, in mountains along the Mexican border. Main range northern and central Mexico.

NOTE Mexican pinyon differs from the common pinyon in its thinner needles in 3's which (in the form present in the United States) lack the whitish lines on the outer surface, in its smaller cones with smaller, hard-shelled seeds or nuts, and in its more southern, limited range along the Mexican border. Though they are widely gathered and eaten in Mexico, Mexican pinyon nuts are of no economic importance in the United States. The small seeds are seldom produced in quantity and are too hard to be cracked with the teeth.

MEXICAN PINYON

MEXICAN PINYON

ADDITIONAL SPECIES *Pinus cembroides* var. *bicolor* Little is sometimes considered a separate species, **border pinyon** (*Pinus discolor* D.K. Bailey & Hawksworth), but as its features intergrade to a large degree with those of *Pinus cembroides,* it is maintained as a variety here.

MEXICAN PINYON

MEXICAN PINYON

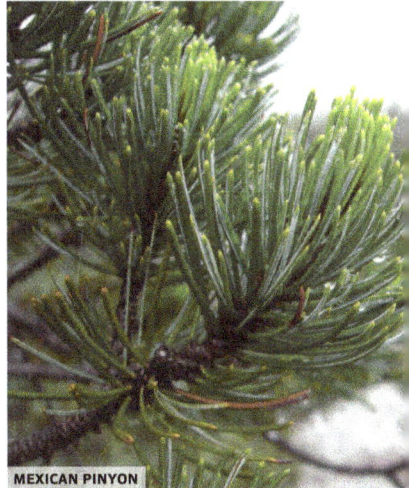

MEXICAN PINYON

Two-Needle Pinyon *Pinus edulis* Engelm.

ALSO CALLED nut pine, pinyon pine, Colorado pinyon pine

SYNONYM *Pinus cembroides* var. *edulis* (Engelm.) Voss

DESCRIPTION Small, bushy, needle-leaf evergreen tree 15 to 35 feet tall, with short trunk 1 to 2 feet or more in diameter, and with compact, rounded, spreading crown.

NEEDLES 2 (sometimes 3 or 1) in a bundle, 1 to 2 inches (2.5 to 5 cm) long, stout, yellow green.

CONES egg-shaped, 1½ to 2 inches (2 to 5 cm) long, light brown, with thick blunt scales and large brown seeds ½ inch (12 mm) long, oily and edible.

BARK furrowed into scaly ridges, gray to reddish brown.

WOOD soft, heavy, resinous, yellow.

HABITAT The abundant and characteristic tree of the pinyon-juniper woodland zone, growing in pure stands or with junipers on dry rocky foothills, mesas, plateaus, and lower mountain slopes between the deserts and forests, mostly from 5,000 to 7,000 feet elevation, widely distributed. Also in northern Mexico.

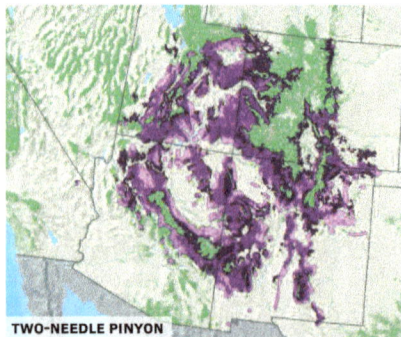

TWO-NEEDLE PINYON - SEEDS

NOTE Pinyon, the state tree of New Mexico, is one of the most abundant and most widely distributed tree species in the Southwest. It grows in pure stands or with one or more of four kinds of junipers and covers vast areas. On the south rim at Grand Canyon National Park it is the commonest tree species. A form with the needles single as in singleleaf pinyon but relatively more slender and shorter occurs in central Arizona along the lower limit of the woodland zone north to Grand Canyon.

TWO-NEEDLE PINYON

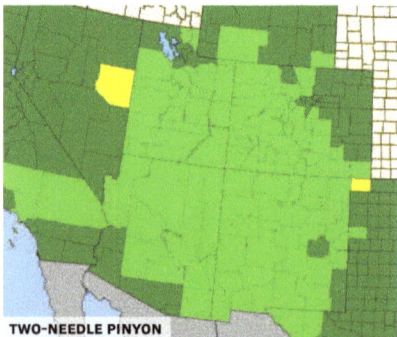

TWO-NEEDLE PINYON

Pinyon is one of the most drought-resistant and slowly growing species of pines, requiring only 12 to 18 inches (30 to 45 cm) of rainfall a year. Because of the scarcity of moisture, these dwarf trees do not form dense forests but grow scattered in open woodlands resembling old orchards.

The familiar seeds, known as pinyon nuts, *piñones*, and Indian nuts, are a wild, commercial nut crop, delicious raw or roasted and also sold shelled and in candies. Pinyon ranks first among the native nut trees of the United States not also under cultivation. Every autumn the Navajo especially harvest large quantities. The pitchy wood is good for fuel and is used locally for mine timbers and fence posts but is not durable unless treated. Pinyons are sometimes planted as ornamentals within their range, where they are hardy but of very slow growth unless irrigated.

TWO-NEEDLE PINYON

Apache Pine *Pinus engelmannii* Carr.

ALSO CALLED Arizona longleaf pine

SYNONYM *Pinus apacheca* Lemmon, *Pinus latifolia* Sarg.

DESCRIPTION Medium-sized, needle-leaved evergreen tree 50 to 70 feet tall and 2 feet or more in trunk diameter, with open, rounded crown and with few large branches.

TWIGS relatively few, stout.

NEEDLES 3 (sometimes 4) in a bundle, 8 to 12 inches (20 to 30 cm) long (to 15 inches (37 cm) in seedlings), stout, dark green.

CONES 4 to 5½ inches (10 to 13 cm) long, light brown, with prickly scales, leaving several scales on the twig when shedding.

BARK deeply furrowed, dark brown.

WOOD hard and heavy, yellowish; similar to that of ponderosa pine but not much used because of the limited supply

HABITAT Uncommon and scattered in pine forest, with Arizona pine, Chihuahua pine, silverleaf oak, and Arizona oak, 5,000 to 8,200 feet elevation, in mountains mostly near the Mexican border; also in adjacent Mexico.

NOTE Apache pine is a distinctive species with a rather restricted range. The seedlings pass through a grasslike stage, reminiscent of longleaf pine (*Pinus palustris*) of the Southeast; they, with the very long-needled saplings, are wholly unlike those of any other western pine. Mature trees, however, are more like the closely related ponderosa pines, differing mainly in the fewer and stouter twigs with longer needles and the slightly larger cones.

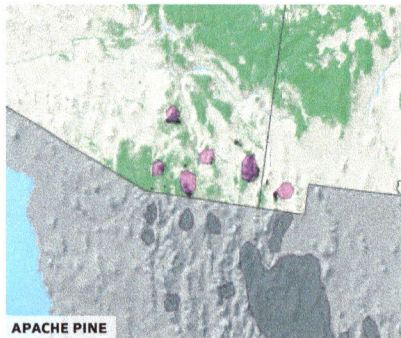

APACHE PINE

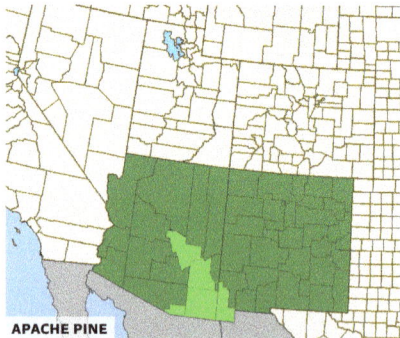

APACHE PINE

APACHE PINE

Limber Pine *Pinus flexilis* James

ALSO CALLED **Rocky Mountain white pine, white pine**

DESCRIPTION Medium-sized, needle-leaved evergreen tree 50 to 80 feet tall and 3 feet or more in trunk diameter, with broad rounded crown.

TWIGS slender, flexible, gray to pale reddish brown, finely hairy when young, soon becoming glabrous.

NEEDLES 5 in a bundle, 2 to 3½ inches (5 to 9 cm) long, slender, blue-green, with fine, white stomatal lines on all surfaces.

CONES short-stalked, hanging down, 4 to 8 inches (10 to 20 cm) long, yellow brown, with thick rounded scales bearing seeds ⅜ to ½ inch (9 to 12 mm) long.

BARK on small trunks smooth and whitish gray, on larger trunks becoming furrowed into broad, rectangular, scaly plates on mature trees.

WOOD soft, lightweight, pale yellow.

HABITAT Not common but widely distributed in high mountains of Douglas-fir and spruce-fir forests (often on exposed ridges and summits), or less frequently in ponderosa pine forest, usually between 7,000 and 10,000 feet elevation (or as low as 6,500 feet elevation in the Mexico border region).

TIP *Pinus strobiformis* is similar, but it has white stomatal lines only on the upper (inner) surface of the needles, and the seed cone scales have recurved tips.

NOTE The thin bark on younger trees of limber pine affords little protection from fires, but the bark on mature trees is much thicker, providing a good degree of protection from low-intensity fires.

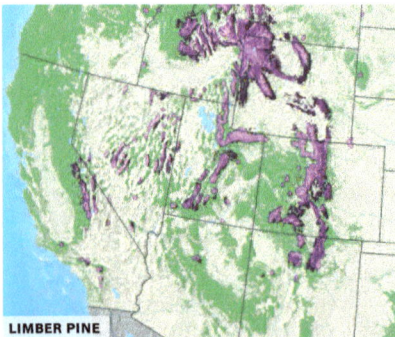

LIMBER PINE

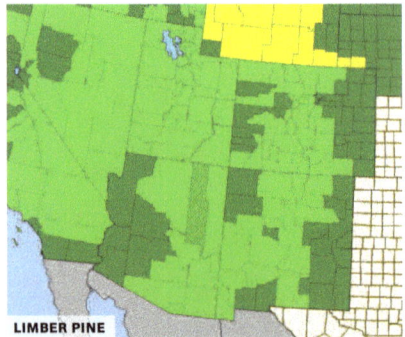

LIMBER PINE

LIMBER PINE

LIMBER PINE

LIMBER PINE

LIMBER PINE

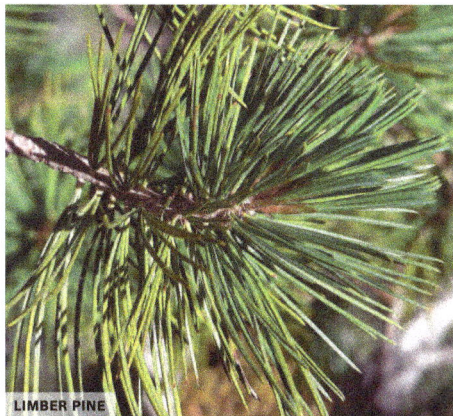

LIMBER PINE

Chihuahuan Pine *Pinus leiophylla* Schiede & Deppe

SYNONYM *Pinus chihuahuana* Engelm.

DESCRIPTION Small to medium-sized tree 30 to 80 feet tall and 1 to 2 feet in trunk diameter, with open, spreading crown. Small, short twigs usually present along the trunk.

NEEDLES 3 in a bundle, 2½ to 4½ inches (6 to 12 cm) long, slender, pale blue green, with the sheath at base soon shedding.

CONES distinctly stalked and spreading, egg-shaped, 1½ to 3 inches (3.5 to 7 cm) long, light brown and shiny, the scales with short, shedding prickles, maturing and opening in 3 years but remaining attached several years.

BARK very thick, 2 inches (5 cm) or more, with deep furrows and broad ridges, dark brown or nearly black.

WOOD hard, orange; because of the small size and limited distribution, of minor importance for lumber.

HABITAT Common or scattered in pine forest, associated with Arizona pine and Apache pine, 5,000 to 7,800 feet elevation, in mountains along the Mexican border; also in Mexico.

NOTE Chihuahuan pine is a northern variety of a widely distributed Mexican species, the typical form having 5 needles in a bundle. This species is easily recognized by the many old, open, stalked cones remaining attached in the branches, by the trunk with very thick bark and often bearing scattered leafy twigs, and by the absence of sheaths at base of needles except when young. This is the only pine native in the USA which requires 3 years instead of 2 to mature its cones. The three stages of developing cones can usually be found on a tree. Chihuahuan pine is one of the few kinds of pine which will sprout from cut stumps.

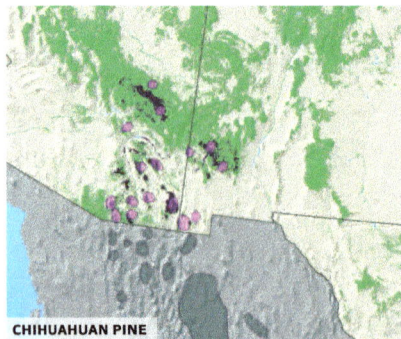

CHIHUAHUAN PINE

CHIHUAHUAN PINE

CHIHUAHUAN PINE

CHIHUAHUAN PINE

CHIHUAHUAN PINE – BARK

Singleleaf Pinyon *Pinus monophylla* Torr. & Frém.

ALSO CALLED nut pine, singleleaf pinyon pine
SYNONYM *Pinus cembroides* var. *monophylla* (Torr. & Frem.) Voss
DESCRIPTION Small, spreading, needle-leaf evergreen tree 15 to 30 feet or more in height, with trunk 1 foot or more in diameter.
NEEDLES 1 in a sheath, 1 to 2 inches (2.5 to 5 cm) long, stout and stiff, gray green.
CONES egg-shaped, 2 to 3 inches (5 to 7.5 cm) long, light brown, with thick blunt scales and large, thin-shelled, brown seeds ¾ inch (2 cm) long, edible, and known as pinyon nuts and pine nuts.
BARK furrowed into scaly ridges, dark brown.
WOOD soft, heavy, yellowish brown; used for firewood.
HABITAT Local in pinyon-juniper woodland with Utah juniper, 4,500 to 6,500 feet elevation. Also in Mexico.
NOTE Singleleaf pinyon, the state tree of Nevada, is a characteristic tree of the pinyon-juniper woodland in the Great Basin region, replacing the species of pinyon in the Southwest.

The nuts of singleleaf pinyon, which are gathered by Native Americans and sold locally on a small scale, differ from the common pinyon nuts of the Southwest in being larger, thinner shelled, and mealy rather than oily in flavor.

SINGLELEAF PINYON

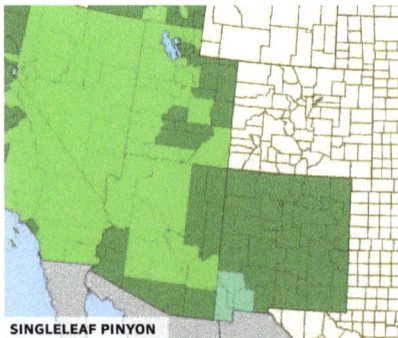

SINGLELEAF PINYON

SINGLELEAF PINYON

SINGLELEAF PINYON

Ponderosa Pine   *Pinus ponderosa* P. & C. Lawson

ALSO CALLED **western yellow pine, yellow pine**

SYNONYM Ours are var. *scopulorum* Engelm.

DESCRIPTION Large, needle-leaved evergreen tree 80 to 125 feet tall with a straight trunk 2 to 3 feet in diameter, or sometimes 150 feet tall and feet in diameter or larger. Crown narrow, with spreading branches.

NEEDLES usually 3 in a bundle or occasionally 2 (mostly 5 in former variety, now species, *Pinus arizonica,* Arizona pine), 4 to 7 inches (10 to 17.5 cm) long, stout, dark green.

CONES 3 to 5 inches (7.5 to 12.5 cm) long, light reddish brown, with prickly scales, leaving several scales on the twig when shedding.

BARK on small trunks less than 1 foot (30 cm) in diameter (known as "blackjacks") blackish and furrowed into ridges, on larger trunks becoming yellow brown and irregularly fissured into large, flat, scaly plates.

WOOD hard, yellowish with whitish sapwood. The lumber has many uses, such as building and other construction, boxes and crates, and millwork; also caskets, furniture, and toys. Other products are piling, poles, fence posts, mine timbers, veneer, and railroad ties.

HABITAT The commonest forest tree in the Southwest and the characteristic species of ponderosa pine forest on the mountains and higher plateaus; in pure stands or associated with Douglas-fir, Gambel oak, limber pine, or pinyon, usually between 5,500 and 8,500 feet in elevation but sometimes beyond these limits; also in northern Mexico.

PONDEROSA PINE

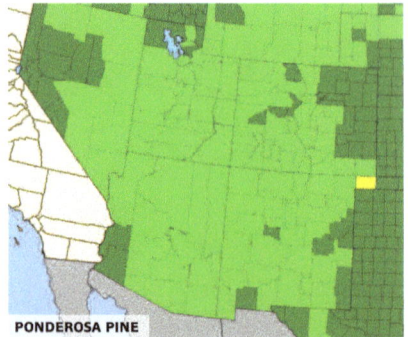

PONDEROSA PINE

NOTE Ponderosa pine has one of the greatest ranges of all the trees in the western mountains. In many places this species grows in pure stands. The vast forest on the Mogollon Plateau in central Arizona (extending into western New Mexico), about 300 miles long, is the largest continuous forest of ponderosa pine anywhere in the West. Ponderosa pine not only is the most valuable saw-timber tree of New Mexico and Arizona but is the most important western pine, second only to Douglas-fir. Most of the lumber produced in the Southwest is from ponderosa pine. The trees are planted also in shelter belts and for ornament. Native Americans stripped and ate the inner bark of this and other conifers. Ponderosa pine is the state tree of Montana.

PONDEROSA PINE

PONDEROSA PINE - BARK

Southwestern White Pine *Pinus strobiformis* Engelm.

ALSO CALLED border limber pine, Mexican white pine

SYNONYM *Pinus flexilis* var. *reflexa* Engelm.

DESCRIPTION Evergreen tree growing to 100 feet high, the trunk to about 3 feet diameter; slender, straight, with conic crown, becoming rounded and irregularly branched, the branches horizontal to drooping. **NEEDLES** in bundles of 5, slender, slightly twisted and flexible, 2½ to 4 inches (6 to 10 cm) long, bright green or blue-green, margins finely sharp-toothed (the teeth very small and widely spaced), to entire; sheaths ⅝ to ¾ inch (1.5-2 cm) long, early deciduous.

CONES mostly 4 to 6 inches (10-15 cm) long, hanging downward, straight to slightly curved, very resinous, yellowish to tan-brown when mature; the scales strongly upturned at their tip; cones on a stalk about ¾ inch (2 cm long), this falling with the cone.

BARK thin, smooth, grayish on young trees, becoming dark grayish red-brown, and divided into small, irregular, rectangular plates.

HABITAT On rocky slopes with other tree species, sometimes in small, pure stands; mid- to high-elevations.

NOTE In northern portions of its range, *Pinus strobiformis* overlaps with (and is thought to hybridize with) **limber pine** (*P. flexilis*). but usually the former has more slender needles. These two can also be distinguished by the needles of *P. strobiformis* being

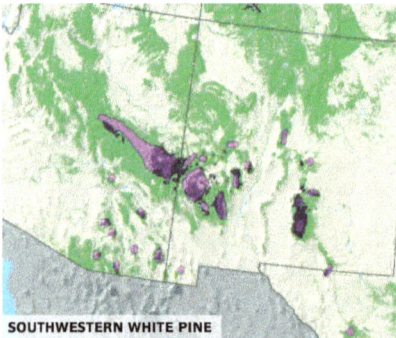

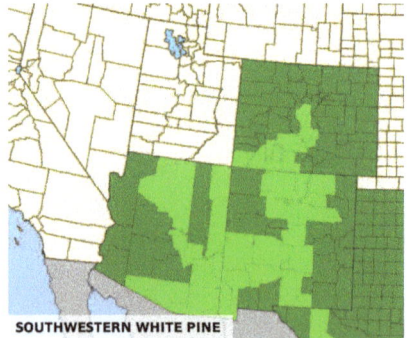

SOUTHWESTERN WHITE PINE

SOUTHWESTERN WHITE PINE

longer, more slender, and bluish green, and the cone scales narrow and strongly reflexed at the tip. In limber pine, the needles are more yellowish-green, and the cone scales are truncate at tip, neither narrowed nor reflexed.

ETYMOLOGY *strobiformis* is from the Greek *strobilos,* meaning twisted, perhaps in reference to the slight twisting of the cone.

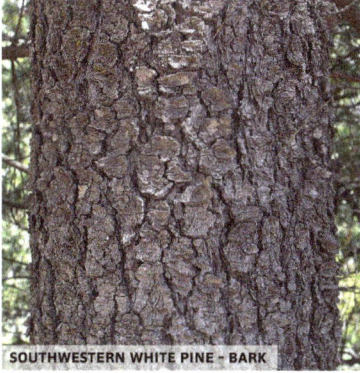

SOUTHWESTERN WHITE PINE - BARK

SOUTHWESTERN WHITE PINE

SOUTHWESTERN WHITE PINE

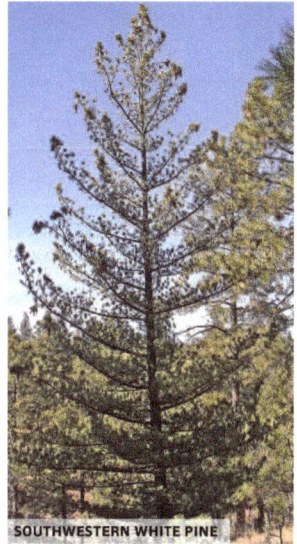

SOUTHWESTERN WHITE PINE

Pseudotsuga DOUGLAS-FIR

Large tree; leaves solitary, short-petiolate, flat, obtuse: cones ovate-oblong, pendulous, the bracts 3-parted, longer than the scales. *Pseudotsuga* from *Pseudo*, false, and *tsuga*, the Japanese word for hemlock.

Douglas-Fir *Pseudotsuga menziesii* (Mirbel) Franco

ALSO CALLED **blue Douglas-fir, Douglas-spruce, red fir, Oregon pine** (lumber)
SYNONYM *Pseudotsuga taxifolia* (Lamb.) Britt.
DESCRIPTION The largest native tree of New Mexico and Arizona, needle-leaved evergreen tree becoming 100 to 150 feet in height, with straight trunk 3 to 6 feet or more in diameter, with open, broadly conical crown and drooping lower branches.
TWIGS slightly roughened by bases of fallen needles.
NEEDLES spreading on all sides of twigs or sometimes in 2 rows, short-stalked, flat, ¾ to 1⅛ inch (2 to 3 cm) long, rounded at tip, dark blue green.
CONES 1¾ to 2¾ inches (4.5 to 7 cm) long, reddish brown, with thin, rounded scales and long, distinctive 3-pointed bracts projecting beyond the scales.
BARK rough and very thick, deeply furrowed into broad ridges, sometimes very corky, dark reddish brown or gray.
WOOD soft, lightweight, yellowish with whitish sapwood.
HABITAT Common as the characteristic tree in the Douglas-fir forest but also occurring in the ponderosa pine forest and spruce-fir forest, 6,500 to 10,000 feet elevation, or as low as 5,200 feet in canyons; widely distributed in higher mountains, also in northern Mexico. Douglas-fir grows in pure stands or mixed with species of the next lower and higher forest zones of the mountains.
NOTE Rocky Mountain trees are classified as var. *glauca* (Beissn.) Franco, which differs from the typical variety of the Pacific Coast region chiefly in its slower growth, more compact habit, shorter and paler blue-green leaves, and smaller cones.

The most important timber tree species of the United States, Douglas-fir ranks first in total stand, lumber production, and production of veneer for plywood. In the Southwest this species is confined to limited areas in the higher mountains but ranks

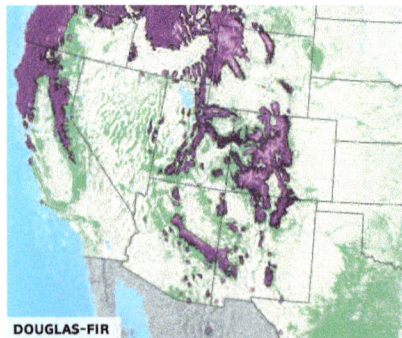

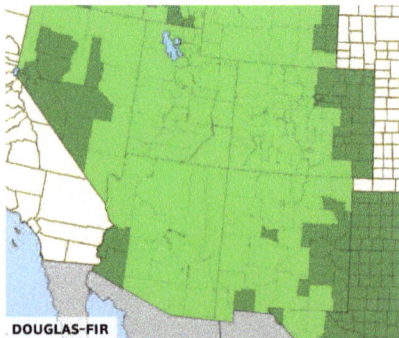

DOUGLAS-FIR

DOUGLAS-FIR

second only to ponderosa pine as a saw timber. Douglas-fir is cut for its high quality lumber in localities where it is sufficiently abundant, chiefly in northern New Mexico and in the Sacramento Mountains of the south-central region of the state. The wood is also used also for rough construction, telephone poles, and railroad ties. Douglas-fir is planted for shade, ornament, and shelter belts and is used as a Christmas tree. It is the state tree of Oregon.

ETYMOLOGY The common name honors *David Douglas* (1798-1834), Scotch botanical explorer, who discovered and introduced to horticulture many species of trees and other plants of the Pacific Northwest; *menziesii* refers to *Archibald Menzies* (1754-1842), a Scottish botanist and surgeon.

DOUGLAS-FIR - CONES

BROADLEAF TREES

ADOXACEAE *Muskroot Family*

Mexican Elder *Sambucus nigra* L.

ALSO CALLED Arizona blueberry elder, Arizona elder, desert elderberry, sauco

SYNONYM *Sambucus mexicana* Presl, *Sambucus glauca* var. *arizonica* Sarg., *Sambucus cerulea* var. *arizonica* Sarg.

DESCRIPTION Evergreen small to medium-sized tree 20 to 30 feet or more in height and 1 to 2 feet in trunk diameter, with compact, rounded crown, and with branches and twigs having large pith.

LEAVES paired, pinnately compound. Leaflets usually 3 (or 5), elliptical or ovate, 1 to 3 inches (2.5 to 7.5 cm) long, short-pointed, finely saw-toothed, thick and leathery, hairy or without hairs.

FLOWER CLUSTER flat-topped, 2 to 8 inches (5 to 25 cm) in diameter, containing many small, yellowish white flowers about ¼ inch across, nearly throughout the year.

BERRIES almost ¼ inch (6 mm) in diameter, dark blue with a bloom, edible.

BARK scaly, light brown.

WOOD soft, lightweight, light brown.

HABITAT Frequent along streams and drainages in desert and desert grassland, 1,200 to 5,000 feet elevation.

MEXICAN ELDER

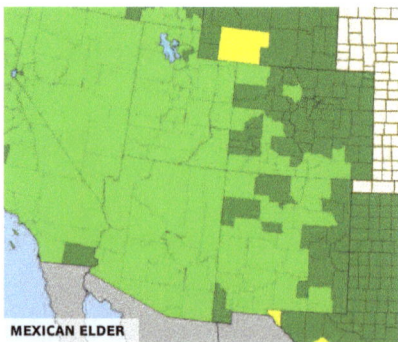

Mexican elder, one of the largest native elders, is distinguished from other southwestern species by its larger size, evergreen leaves, and occurrence in the desert, rather than mountains. These evergreen trees with numerous showy flower clusters are often cultivated as ornamentals and shade trees. The fruits are used in pies and jellies and are eaten by birds.

NAME *Sambucus* is the Latin name, from *sambuca,* a kind of harp perhaps made of elder wood.

MEXICAN ELDER

ANACARDIACEAE *Cashew Family*

Rhus SUMAC

Shrubs, the sap usually acrid and resinous; leaves alternate, either simple and entire, or 3-foliolate (sometimes with 5 leaflets); flowers regular, perfect or unisexual, mostly 5-merous, small, greenish, yellowish, or whitish, in axillary or terminal panicles, with a ring-shaped or cup-shaped disk around the ovary; fruit a small 1-seeded drupe. The Greek name for one species of *Rhus*.

1   Leaves compound . . . . MEARNS' SUMAC (*Rhus virens*)
1   Leaves simple, evergreen, leathery, entire (or very nearly so) 2

2   Blades of the leaves broadly ovate, acute or short-acuminate, often somewhat folded, bright green above; petioles usually more than ⅜ inch (1 cm) long; northern and central Arizona. . . . . SUGAR SUMAC (*Rhus ovata*)
2   Blades broadly oblong or oval, obtuse or acutish, flat, dark green above with conspicuous whitish veins; petioles usually less than ⅜ inch (1 cm) long; rare in Yuma County, Arizona. . . . . KEARNEY'S SUMAC (*Rhus kearneyi*)

Kearney's Sumac *Rhus kearneyi* F.A. Barkley

DESCRIPTION Evergreen large shrub or small tree to 18 feet in height.
LEAVES oblong, 1 to 2¼ inches (2.5 to 5.5 cm) long, rounded or blunt-pointed at tip, edges without teeth but rolled under, very leathery, almost without hairs.
FLOWERS small in short, crowded clusters at tips of twigs, less than ¼ inch (6 mm) across, whitish.
FRUIT oblong, ⅜ inch (9 mm) long, slightly flattened, reddish and hairy.
HABITAT Known only from Tinajas Altas Mountains, southern Yuma County, where this species was discovered on dry slopes in desert, 1,000 to 1,500 feet elevation.
NAMED FOR Thomas H. Kearney, botanist of the US Dept. of Agriculture and co-author of *Flowering Plants and Ferns of Arizona.*

KEARNEY'S SUMAC

KEARNEY'S SUMAC

Sugar Sumac *Rhus ovata* S. Wats.

ALSO CALLED sugarbush, mountain-laurel

DESCRIPTION Evergreen shrub or small tree to 15 feet in height and 5 inches in trunk diameter, with rounded crown.

LEAVES ovate 1½ to 3¼ inches (3.5 to 8 cm) long, short-pointed, rounded at base, edges without teeth, thick and leathery, not flat but curved upward at midrib, shiny light green on both sides, hairless.

FLOWERS small in crowded clusters 2 inches (5 cm) long at tips of twigs, about ¼ inch (6 mm) across, pink or reddish in buds but cream-colored when open, in April.

FRUIT reddish, hairy, sweetish.

BARK rough, shaggy and very scaly, gray brown.

WOOD light brown.

HABITAT Common on mountain slopes in chaparral, 3,000 to 5,000 feet elevation; not in New Mexico.

NOTE This attractive broadleaf evergreen is one of several shrubby species of the chaparral vegetation in southern California which reappear in the same zone in the mountains of central Arizona. In California it is planted for erosion control and landscaping in mountains and as an ornamental. The common name refers to the sweetish, edible fruit coats, which were used as sugar by the Native Americans. The broad, shiny, evergreen leaves, reddish twigs, and sticky drupes help to distinguish this species.

SUGAR SUMAC

SUGAR SUMAC

Mearns' Sumac *Rhus virens* Lindheimer ex Gray

SYNONYM *Rhus choriophylla* Woot. & Standl.

DESCRIPTION Evergreen shrub usually less than 7 feet tall, rarely a small tree 15 feet in height and 3 inches (7.5 cm) in trunk diameter, with relatively few branches.

LEAVES pinnately compound, 2 to 3 inches (5 to 7.5 cm) long. Leaflets 3 or 5, ovate, 1 to 2 inches (2.5 to 5 cm) long, short-pointed, edges without teeth, thick and leathery, shiny green above and dull green below, slightly hairy.

FLOWERS small, several in cluster, about 1/8 inch (3 mm) long, whitish, in August and September.

FRUIT ¼ inch (6 mm) long, reddish, hairy.

BARK scaly and shaggy, gray.

HABITAT Dry rocky slopes especially on limestone, canyons and mountains in oak woodland, 4,000 to 6,000 feet elevation. Previously known only as a shrub, Mearns sumac has been reported as a small tree in Santa Cruz County, Arizona.

NOTE A large evergreen shrub distinguished by its pinnate leaves with 5-9 large, leathery, shiny leaflets with pointed tips; and the white flowers in clusters followed by characteristic red, sour-tasting sumac berries.

ETYMOLOGY This species was discovered by *Edgar A. Mearns* (1856-1916), American naturalist and Army surgeon, who made large plant and animal collections along the Mexican boundary from 1891 to 1894.

MEARNS' SUMAC

MEARNS' SUMAC

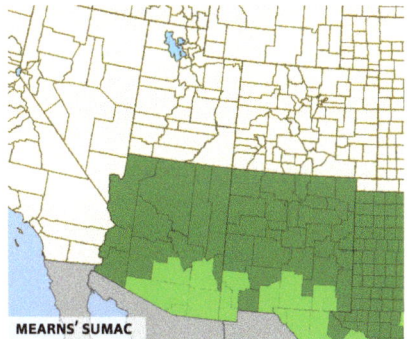

MEARNS' SUMAC

BETULACEAE *Birch Family*

Trees or large shrubs; leaves simple; flowers monoecious, appearing with or before the leaves, those of both sexes in catkins, the staminate catkins pendulous, the pistillate flowers subtended by conspicuous bracts; ovary 2-celled; fruit a 1-seeded nutlet.

1 Nutlets wingless, each enclosed in a large, bladderlike, papery bract; staminate flower solitary in the axil of each scale .... **HOP-HORNBEAM** (*Ostrya*)
1 Nutlets winged, not enclosed in a bladderlike bract; staminate flowers more than one in the axil of each scale **2**

2 Pistillate catkins solitary, their scales remaining thin, deciduous with or soon after the nutlets, not wedge-shaped, deeply 3-lobed, the midlobe elongate ..... **BIRCH** (*Betula*)
2 Pistillate catkins usually several in a raceme-like cluster, their scales becoming thick and woody, long-persistent on the branch after the nutlets have fallen, wedge-shaped, shallowly 3- to 5-lobed .... **ALDER** (*Alnus*)

Alnus ALDER

Shrubs or small trees with thin toothed leaves; sterile catkins with 4 or 5 bractlets and 3 flowers upon each scale; fertile catkins ovoid or ellipsoid, the scales each subtending 2 flowers and a group of 4 small scales, the latter becoming woody in fruit. *Alnus* is the classical Latin name.

1 Shrub or small tree; leaf blades deeply and doubly serrate-dentate, often somewhat lobed; stamens 4................................. **THINLEAF ALDER** (*Alnus incana*)
1 Medium to large tree; leaf blades shallowly and doubly serrate-dentate, seldom lobed; stamens 1 to 3 (usually 2) .... **ARIZONA ALDER** (*Alnus oblongifolia*)

Thinleaf Alder *Alnus incana* (L.) Moench

ALSO CALLED mountain alder, speckled alder
SYNONYM *Alnus tenuifolia* Nutt.
DESCRIPTION Large shrub or small tree to 30 feet tall, with usually several trunks up to 6 inches (15 cm) or more in diameter spreading from the base, and with rounded crown.
LEAVES ovate or oblong, 2 to 4 inches (5 to 10 cm) long, short-pointed at tip, rounded, straight, or slightly heart-shaped at base, edges doubly saw-toothed and slightly lobed, thin, dark green and hairless above, beneath pale yellow green and hairless or slightly hairy.
MALE AND FEMALE FLOWERS in catkins in early spring.
FRUIT of cones ⅜ to ½ inch (9 to 12 mm) long, with hard black scales and many small nutlets, remaining on tree in winter.
BARK on small trunks, thin and grayish but becoming scaly and reddish brown.
WOOD light brown.

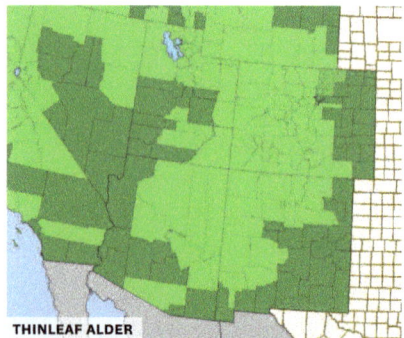

THINLEAF ALDER

HABITAT Along streams and canyons mostly in mountains (often forming thickets on streambanks), ponderosa pine forest, 7,000 to 9,000 feet elevation.

NOTE The Navajos made a red dye for wool from the powdered bark of this species, together with ashes of one-seed juniper and a decoction of mountain-mahogany.

THINLEAF ALDER

THINLEAF ALDER

THINLEAF ALDER

Arizona Alder *Alnus oblongifolia* Torr.

ALSO CALLED New Mexican alder

DESCRIPTION Medium-sized to large tree to 60 or 80 feet in height, with tall straight trunk 2 to 3 feet in diameter, and with open, rounded crown.

LEAVES elliptic, 2 to 3 inches (5 to 7.5 cm) long, short-pointed at tip, gradually narrowed at base, edges sharply and usually doubly saw-toothed, thin, dark green and almost hairless above, beneath paler and slightly hairy.

MALE AND FEMALE FLOWERS in catkins in March.

FRUIT of cones about ½ inch (12 mm) long, with hard, black scales and many small nutlets, remaining on tree in winter.

BARK smooth, thin, dark gray, on large trunks fissured and scaly.

WOOD soft, lightweight, whitish when cut but turning to light reddish brown.

HABITAT Along canyons and streams mostly in mountains, oak woodland and ponderosa pine forest, 4,500 to 7,500 feet elevation.

NOTE The doubly serrate leaves are only found in our region in the Birch Family, while our only other alder, *Alnus incana,* is usually a shrub, has leaves that are more rounded and lobed at their base, and is generally found at higher elevations in the mountains.

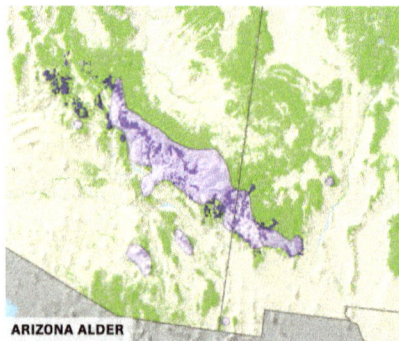

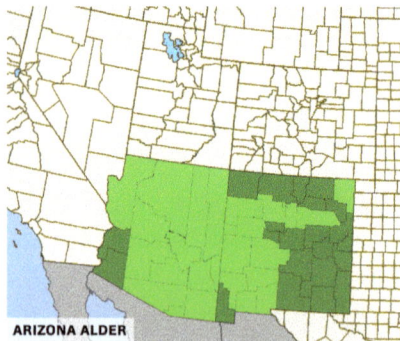

ARIZONA ALDER

ARIZONA ALDER

Water Birch *Betula occidentalis* Hook.

ALSO CALLED **red birch, black birch**

SYNONYM *Betula fontinalis* Sarg.

DESCRIPTION Small tree or large shrub to 25 feet in height and 8 inches in trunk diameter, with finely branched, spreading, open crown.

TWIGS often drooping, slender, covered with resinous glands, light green when young but becoming dark red brown.

LEAVES ovate, 1 to 2 inches (2.5 to 5 cm) long, short or long-pointed, usually rounded at base, sharply and often doubly saw-toothed, thin, dark green above, beneath pale yellow green and shiny and covered with resin dots.

MALE AND FEMALE FLOWERS in catkins in early spring.

FRUIT of erect or hanging cylindrical cones 1 to 1¼ inches (2.5 to 3 cm) long and ½ inch (12 mm) thick, slightly hairy, with many small winged nutlets.

BARK smooth, shiny, copper colored, with conspicuous pale brown horizontal lines (lenticels).

WOOD soft, light brown.

HABITAT Along streams in mountains, often forming thickets, pinyon-juniper woodland and ponderosa pine forest zones, 5,000 to 8,000 feet elevation.

NOTE Water birch is the only native birch in the Southwest.

ETYMOLOGY *Betula* is the Latin name.

WATER BIRCH

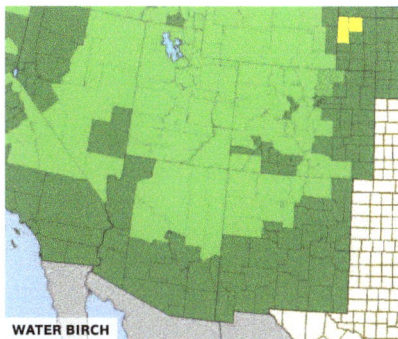

WATER BIRCH

Knowlton's Hop-Hornbeam *Ostrya knowltoni* Sarg.

ALSO CALLED **western hophornbeam**

SYNONYM *Ostrya baileyi* Rose

From the Greek name for some other hardwood tree.

DESCRIPTION Shrub or small slender tree 10 to 40 feet tall and 6 inches (15 cm) in trunk diameter.

YOUNG TWIGS brown, hairy and with glandular hairs.

LEAVES elliptic to ovate, 1 to 2½ inches (2.5 to 6 cm) long, short-pointed or rounded at tip, usually rounded or slightly heart-shaped at base, edges sharply and doubly saw-toothed, dark yellow green and slightly hairy above, paler and soft hairy beneath.

MALE AND FEMALE FLOWERS in catkins in April.

FRUIT conelike, 1 to 1½ inches (2.5 to 3.5 cm) long and ¾ inch (2 cm) broad, of baglike, papery bracts, each enclosing a nutlet ¼ inch (6 mm) long.

BARK thin, fissured and peeling off in long flakes, gray.

WOOD hard, light reddish brown.

HABITAT Very local in moist canyons, oak woodland, pinyon-juniper woodland, and lower ponderosa pine forest zones, 4,200 to 7,000 feet elevation. Common below both rims of the Grand Canyon.

NOTE As the common name suggests, the bladdery fruit resembles that of hops. Knowlton hophornbeam is the only representative of its genus in western United States but another species, **eastern hop-hornbeam** (*Ostrya virginiana*), is widely distributed in the eastern half of the country. The plants of southeastern New Mexico and adjacent Texas are slightly intermediate between the eastern and western species.

KEY CHARACTERS *Ostrya knowltonii* is similar to **Water Birch** (*Betula occidentalis*), but it has inflated, bladder-like bracts surrounding each nutlet

ETYMOLOGY *Frank H. Knowlton* (1869-1926), American botanist and paleobotanist of the United States Geological Survey, discovered this species at the Grand Canyon in 1889.

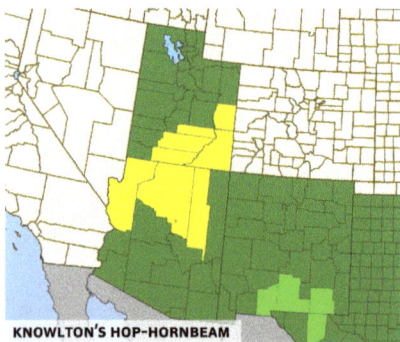

KNOWLTON'S HOP-HORNBEAM

BIGNONIACEAE *Bignonia Family*

Desert-Willow *Chilopsis linearis* (Cav.) Sweet

DESCRIPTION Large shrub or small tree to 25 feet or more in height, with spreading crown.

TWIGS slender, varying in varieties from woolly to sticky or neither.

LEAVES very narrow or narrowly lance-shaped, 3 to 6 inches (7.5 to 15 cm) long and less than ⅜ inch (9 mm) wide, straight or curved, not toothed, light green.

FLOWERS several in clusters, large and showy, tubular, 1 to 1¼ inches (2.5 to 3 cm) long, whitish and tinged with purple or pink, fragrant, from April to August.

SEED CAPSULES long and narrow, 4 to 8 inches (10 to 20 cm) long and less than ¼ inch (6 mm) in diameter, remaining attached in winter, containing many flat seeds with 2 hairy wings.

BARK ridged and scaly, dark brown.

WOOD soft, brown streaked with yellow; durable and suitable for fence posts.

HABITAT Along washes and drainages in plains, mesas, and foothills, desert and desert grassland, 1,500 to 5,000 (or 6,000) feet elevation.

NOTE In spite of its general aspect and habitat in moist places, desert-willow is not a true willow. Instead, it belongs to the same family as the catalpa tree, and has similar large flowers and long narrow fruits. It forms thickets along washes, and is important for erosion control, and has been widely planted for this purpose. It is propagated from seeds or cuttings, grows rapidly, and sprouts after being cut. Desert-willow is sometimes used as an ornamental. Arizona plants are subsp. *arcuata,* which is characterized by its strongly arcuate (curved into an arch-shape) leaves. New Mexico plants are both subsp. *arcuata* as well as subsp. *linearis,* which has straight or only slightly curved leaves.

KEY CHARACTERS are the long slender leaves; long, slender seed pod which releases seeds that have tufts of hairs on both ends; and the attractive, bilabiate, pink to lavender flowers.

ETYMOLOGY Gr. *cheilos,* a lip; *opsis,* resembling. The calyx of this low tree or shrub has a distinct lip.

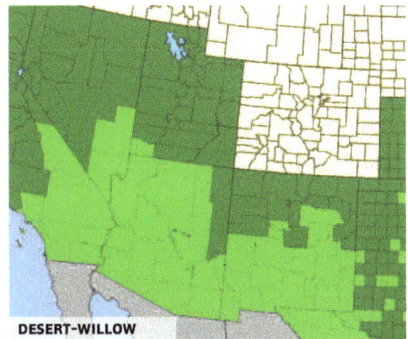

DESERT-WILLOW

DESERT-WILLOW

DESERT-WILLOW

BURSERACEAE *Bursera Family*

Bursera BURSERA

Shrubs or small trees, unarmed, strongly aromatic; young bark smooth and brown, the older bark exfoliating; leaves alternate, pinnate, deciduous; flowers small, solitary or in very few-flowered clusters; petals inserted on a ring-shaped disk; fruit drupelike, 3-angled, with 1 large bony seed.

A resin called copal is obtained in Mexico from many of the species, including *B. fagaroides.* It is used for cement and varnish, and for treating bites of scorpions, etc. It is burned as incense in churches and was formerly so employed in the Aztec and Mayan temples. Bursera is named for the Danish botanist *Joachim Burser* (1583-1639).

1　Leaflets lanceolate, acute or acutish at tip, ⅝ to 1¾ inches (15 to 40 mm) long, ³⁄₁₆ to ⅝ inches (4 to 15 mm) wide; inflorescence several-flowered; pedicels often borne on a common peduncle as long as or longer than themselves; young bark gray brown, the old bark exfoliating in large thin sheets . . . **FRAGRANT BURSERA** (*Bursera fagaroides*)

1　Leaflets narrowly oblong, oval, or spatulate (the terminal one sometimes nearly orbicular), obtuse at tip, ¹⁄₁₆ to ⅜ inches (2 to 9 mm) long, ¹⁄₁₆ to ⅛ inches (1 to 3 mm) wide; inflorescence very few-flowered; pedicels with or without a common peduncle; young bark red brown, the old bark exfoliating in flakes . . . . **ELEPHANT-TREE** . . . . (*Bursera microphylla*)

Fragrant Bursera *Bursera fagaroides* (Kunth) Engl.

SYNONYM *Bursera odorata* T. S. Brandegee
DESCRIPTION Strongly aromatic shrub or small tree to 15 feet tall.
LEAVES pinnately compound, 2 to 4 inches (5 to 10 cm) long, with winged axis, aromatic, with odor of tangerine peel when crushed. Leaflets 5 to 11, lance-shaped, ½ to 1½ inches (12-35 mm) long, edges without teeth, hairless.
FLOWERS several in a cluster below a group of new leaves at end of a short side twig, in July.
FRUITS 3-angled, less than ⅜ inch (9 mm) long, gray, 1-seeded, splitting into 3 parts.
BARK of branches gray brown, on old trunks peeling off in large, thin, papery sheets.
HABITAT Rare (and now likely extirpated) at one Arizona locality (Baboquivari Mountains, Pima County) near Mexican border, on dry limestone cliffs in desert mountains, 4,000 feet elevation; not in New Mexico; also in Mexico.
NOTE The young bark is bright green, and the crushed herbage has an odor of tangerine peel. The old bark can be removed in translucent sheets resembling parchment. The branches yield quantities of gum, and when cut from the tree dry very slowly.

FRAGRANT BURSERA (likely extirpated from Arizona)

FRAGRANT BURSERA

FRAGRANT BURSERA

Elephant-Tree *Bursera microphylla* A. Gray

ALSO CALLED torote, copal, elephant bursera

DESCRIPTION Strongly aromatic shrub or small tree to 20 feet tall, with relatively thick, swollen trunk to 1 foot in diameter, short and sharply tapering, with stout, crooked, tapering branches, and with widely spreading, rounded, open crown of thin foliage.

TWIGS reddish brown, hairless.

LEAVES pinnately compound, 1 to 1¼ inches (2.5 to 3 cm) long, with winged axis, aromatic. Leaflets 15 to 30, small, narrowly oblong leaflets about inch long, edges without teeth, hairless.

FLOWERS single or 2 or 3 at a node, short-stalked, small, less than ¼ inch (6 mm) long, whitish, partly of separate sexes, in July.

FRUITS 3-angled, ¼ inch (6 mm) long, red, 1-seeded, splitting into 3 parts, aromatic.

BARK of trunk and branches papery and peeling off in thin flakes, white on outside, the next thin layers green, and the inner thick layers red and corky.

WOOD hard, pale yellow.

HABITAT Common locally on rocky slopes of very arid desert mountains, 1,000 to 2,500 feet elevation; not in New Mexico.

NOTE The common name is from the stout, tapering branches which resemble an elephant's trunk. The trees are not resistant to cold and are killed back when young. The bark contains tannin and was gathered in Sonora for export; in that region the gum was used for treating venereal diseases.

ELEPHANT-TREE

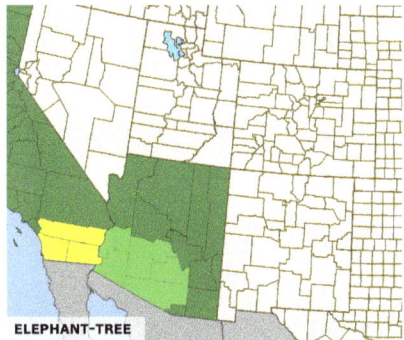

ELEPHANT-TREE

ELEPHANT-TREE

CANNABACEAE *Hemp Family*

Netleaf Hackberry *Celtis reticulata* Torr.

ALSO CALLED **paloblanco, western hackberry**
SYNONYM *Celtis laevigata* var. *reticulata* (Torr.) L. Benson
ETYMOLOGY Greek name for another tree.
DESCRIPTION Large shrub or small tree to
30 feet tall and 1 foot or more in trunk
diameter, with spreading crown.
LEAVES in 2 rows, very variable, mostly
ovate, 1 to 2½ inches (2.5 to 6 cm) long,
one-sided, shortor long-pointed, base
rounded or slightly heart-shaped and with
3 main veins, edges without teeth or
sometimes coarsely saw-toothed, usually
thick, dark green and rough above,
beneath yellow-green, strongly veined, and
slightly hairy, shedding in autumn or
winter.
FLOWERS small, greenish, in March and
April.
FRUIT ¼ to ⅜ inch (6 to 9 mm) in diameter,
orange-red, dry and sweet, 1-seeded.
BARK smoothish or becoming rough and
fissured, with large corky warts, gray.
WOOD light brow, used for fuel and fence
posts
HABITAT Usually along streams, canyons,
and washes, in moist soil, plains grassland,
upper desert, desert grassland, and oak
woodland zones, 2,500 to 6,000 feet
elevation, widely distributed.
NOTE The sweetish fruits are eaten by
wildlife. Leaves often have rounded, swollen
galls caused by insects, while the branches
often have witches' broom galls.

NETLEAF HACKBERRY

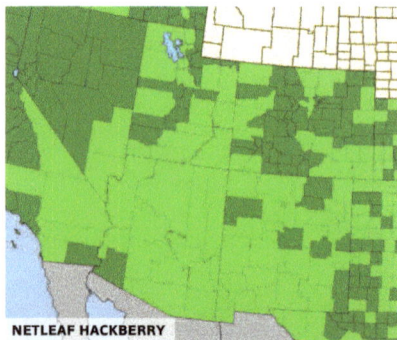

NETLEAF HACKBERRY

NETLEAF HACKBERRY

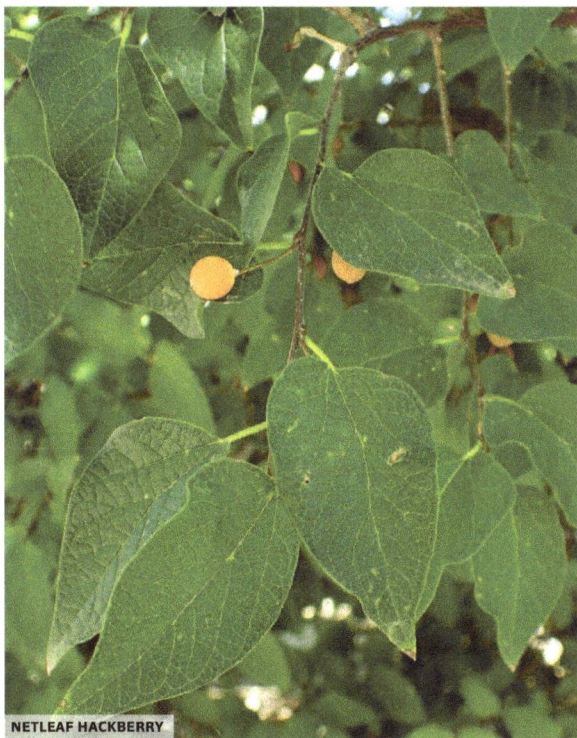

NETLEAF HACKBERRY

CELASTRACEAE *Bittersweet Family*

Canotia *Canotia holacantha* Torr.

ALSO CALLED crucifixion-thorn

DESCRIPTION Spiny shrub or small tree to 18 feet in height, with short trunk up to 1 foot in diameter, with many flexible, upright, yellowish green, rushlike branches and twigs in broomlike masses, leafless most of the year and resembling paloverdes.

TWIGS ¹⁄₁₆ to ⅛ inch (1.5 to 3 mm) in diameter, inconspicuously grooved, hairless, with small black rings at base where forked, partly ending in spines or dead tips.

FLOWERS 3 to 7 in small clusters along the twigs, nearly ¼ inch (6 mm) in diameter, greenish white, May to August.

FRUIT a hard, egg-shaped, long-pointed capsule, ¾ inch (2 cm) or more in length, reddish brown, 5-celled and splitting open along 10 lines; seeds single or paired and winged.

BARK on branches smooth and yellow green, at base fissured, becoming rough and slightly shreddy, gray.

HABITAT Common or abundant locally on dry slopes and hillsides in upper desert and lower chaparral, 2,000 to 5,000 feet elevation; not in New Mexico.

NOTE Canotia resembles the paloverdes in its yellowish green, leafless branches and twigs which are more crowded, upright, and in broomlike masses. It replaces paloverdes as a common desert tree in the northern part of its range, and is the commonest of the three spiny, much-branched shrubby species called "crucifixion-thorns."

ETYMOLOGY *Canotia* is a nineteenth century Mexican name for these plants.

CANOTIA

CANOTIA

CANOTIA

CORNACEAE *Dogwood Family*

Red-Osier Dogwood *Cornus sericea* L.

SYNONYM *Cornus alba* L., Cornus *stolonifera* Michx.

DESCRIPTION Usually a shrub 8 feet or less in height but reported as rarely a small tree in southern Arizona.

BRANCHES AND TWIGS purplish red, finely hairy.

LEAVES paired, ovate or elliptical, 2 to 3½ inches (5 to 9 cm) long, short-pointed, edges not toothed, dark green and nearly hairless above, beneath pale or whitish, and finely hairy, with about 5 long and curved veins on each side of midrib.

FLOWERS several in flat-topped cluster 1 to 2 inches (2.5 to 5 cm) broad, small, less than ¼ inch (6 mm) across, white, in June and July.

FRUIT white, ¼ inch (6 mm) in diameter, 1-seeded.

BARK greenish brown, smooth.

HABITAT Common along streams and canyons in mountains and plateaus, ponderosa pine forest and Douglas-fir forest, 6,000 to 9,000 feet elevation, widely distributed.

NOTE Red-osier dogwood is reported to reach tree size in the Santa Catalina Mountains in southern Arizona. The common name refers to the resemblance of the reddish branches and twigs to those of some willows.

ETYMOLOGY *Cornus* is the Latin name for the cornelian cherry (*Cornus mas*).

RED-OSIER DOGWOOD

RED-OSIER DOGWOOD

ERICACEAE *Heath Family*

Arbutus MADRONE

Small to medium trees with exfoliating bark; leaves evergreen, leathery, alternate, petiolate; flowers small, white or flesh-colored, in a terminal cluster of racemes or panicles; calyx small, 5-parted; corolla globular to ovoid; ovules crowded on a fleshy placenta projecting from the inner angle of each cell; styles long, the stigmas obtuse. *Arbutus* is the Latin name.

1   Leaves elliptic-lanceolate, acute, glabrous; southern Arizona and southwestern New Mexico . . . . ARIZONA MADRONE *Arbutus arizonica*
1   Leaves oblong or ovate, obtuse, permanently pubescent beneath; southeastern New Mexico . . . . TEXAS MADRONE *Arbutus xalapensis*

Arizona Madrone *Arbutus arizonica* (A. Gray) Sarg.

ALSO CALLED Arizona madrono
SYNONYM *Arbutus xalapensis* var. *arizonica* Gray.
DESCRIPTION Evergreen medium-sized tree to 50 feet or more in height and 2 feet or more in trunk diameter, with compact, rounded crown.
TWIGS finely hairy, becoming reddish brown and peeling off in scales.
LEAVES lance-shaped, 1½ to 3 inches (3.5 to 7 cm) long, short-pointed, edges without teeth or inconspicuously saw-toothed, thick and rigid, shiny light green above and paler beneath.
FLOWERS in hairy terminal clusters 2 to 2½ inches (5 to 6 cm) long, urn-shaped, ¼ inch (6 mm) long, white or pink, from April to September.
FRUIT berrylike, ⅜ inch (9 mm) in diameter, orange red, finely warty, mealy and sweetish, with large stones.
BARK of branches thin, reddish brown, peeling off in thin scales, on larger branches and trunks divided into squarish plates and light gray or whitish.
WOOD heavy, light brown tinged with red.
HABITAT Mountains in oak woodland, 4,000 to 8,000 feet elevation, near Mexican border; also in northern Mexico.
NOTE Arizona madrone should make an attractive ornamental tree, though probably of slow growth. Under some conditions, drops of water exude from the leaves, as if the trees were weeping or were "raintrees."

ARIZONA MADRONE

ARIZONA MADRONE

TEXAS MADRONE

Texas Madrone *Arbutus xalapensis* Kunth

ALSO CALLED **Texas madrono**

SYNONYM *Arbutus texana* Buckl.

DESCRIPTION Evergreen small tree to 20 feet tall and 1 foot in trunk diameter.

TWIGS bright red, densely hairy when young, becoming dark red brown and scaly.

LEAVES oval or lance-shaped, 1 to 3 inches (2.5 to 7.5 cm) long, rounded or short-pointed at tip, edges entire or small-toothed, thick and rigid, dark green and hairless above, paler and slightly hairy beneath.

FLOWERS in hairy terminal clusters 2½ inches (6 cm) long, urn-shaped, ¼ inch (6 mm) long, white or pink.

FRUIT berrylike, ⅜ inch (9 mm) in diameter, dark red, finely warty, mealy and sweetish, with large stone.

BARK of branches thin, orange-red, separating into papery scales, on larger trunks divided into square plates and dark reddish brown.

WOOD hard, heavy, brown tinged with red.

HABITAT Local in mountains in oak woodland zone, about 6,000 feet elevation; not in Arizona.

NOTE Similar to *Arbutus arizonica* but the ovary in that species is glabrous rather than hairy as in *A. xalapensis;* see key for other differences.

TEXAS MADRONE

TEXAS MADRONE

EUPHORBIACEAE *Spurge Family*

Herbs, shrubs, or trees, often with milky juice (in *Sebastiana*); leaves simple (*Sebastiana*) or compound (*Ricinus*); flowers either male or female on same plant; fruit a capsule.

1 Leaves very large, palmately lobed; plants not with mily sap . . . . **CASTOR-BEAN**
. . . . (*Ricinus communis*)
1 Leaves smaller, entire, not palmately lobed; plants with mily sap . . . .
. . . . **JUMPING-BEAN SAPIUM** (*Sebastiania bilocularis*)

Castor-Bean *Ricinus communis* L.

DESCRIPTION Herb, shrub, or in warm regions a small tree to 15 feet or more in height, with hollow stems and leaf-stalks.
LEAVES very long-stalked, very large, nearly circular, 1 foot or more in diameter, with usually 7 to 10 fingerlike (palmate) lobes surrounding the leafstalk, edges saw-toothed, thick, often reddish or purplish tinged.
FLOWERS in large branched clusters at tops of branches, female flowers above and male flowers below.
SEED CAPSULE ⅝ to 1 inch (15 to 25 mm) in diameter, spiny, 3-celled, splitting open, with 3 large, poisonous seeds ½ to ¾ inch (12 to 20 mm) long.
BARK smooth, light gray.
HABITAT Escaping from cultivation and naturalized along washes and streams in desert in Arizona (but not New Mexico). Probably native of Africa but widely cultivated and naturalized over the world, escaping from cultivation in warmer parts of the United States.
NOTE Castor-bean is planted for ornament, as a large annual herb in temperate regions, and as a perennial shrub or small tree in regions with mild winters. The raw seeds are very poisonous. It is reported that eating as few as three seeds may cause death. Other parts of the plant also contain the toxic principle, which is destroyed by heat. Castor oil, extracted from the seeds, is used in medicine and as a lubricant.
KEY FEATURES (1) Leaves large, peltate, palmately lobed; (2) juice not milky; (3) pods spiny, borne at the tips of the branches; (4) stamens extremely numerous (in the staminate flowers).
ETYMOLOGY Latin, *ricinus,* a tick; from the resemblance of the seeds.

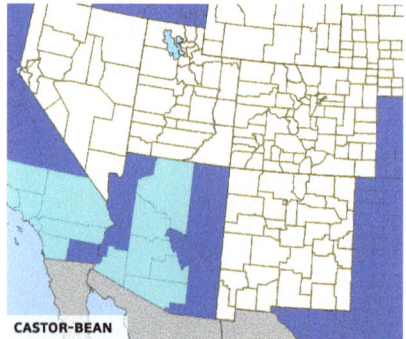

CASTOR-BEAN

Jumping-Bean Sapium *Sebastiania bilocularis* S. Watson

ALSO CALLED Mexican jumping-bean

SYNONYM *Sapium biloculare* (S. Wats.) Pax

DESCRIPTION Tall shrub or small tree to 15 feet in height, with milky juice.

LEAVES lance-shaped, ¾ to 2¼ inches (12 to 55 mm) long, long-pointed, edges finely saw-toothed, leathery, without hairs.

FLOWERS in narrow zigzag clusters 1 to 2¼ inches (2.5 to 5.5 cm) long at ends of twigs, stalkless, fragrant male flowers above and 1 or 2 female flowers at base, from March to November.

SEED CAPSULE 2-lobed, nearly ½ inch (12 mm) broad, splitting open, with 2 nearly spherical, mottled brownish gray seeds about ½ inch (12 mm) long.

HABITAT Common locally in rocky slopes and sandy washes of desert, 800 to 2,500 feet elevation; not in New Mexico.

NOTE This species produces 'Mexican jumping-beans,' though it is not the important, widely distributed Mexican jumping-bean. The larva of a small moth, which often infests the seeds, by its movements causes a seed to move and roll slightly. The milky juice is poisonous. It has been used as a fish poison and by Native Americans to poison their arrows. A very small amount of this juice rubbed into the eyes accidentally after a plant is handled will cause extreme pain for hours.

ETYMOLOGY The former name, *Sapium,* is the Latin name for a resinous pine; the stems of these shrubs or trees, which are not pines, exude a sticky sap.

JUMPING-BEAN SAPIUM

JUMPING-BEAN SAPIUM

FABACEAE *Pea Family*

Herbs, shrubs, or small trees; leaves alternate, mostly compound, but simple in 2 of our species; flowers mostly perfect, commonly irregular, the petals separate or partly united (especially the 2 lowest or keel petals in papilionaceous flowers), commonly 5, rarely only 1 or none; fruit a 2-valved pod.

This important family contains about 500 genera and over 15,000 species among which are many important food plants. The family is characterized by its distinctive pod-like fruit known as a legume; as well as by alternate, usually compound leaves; and regular or distinctive papilionaceous flowers.

1 Leaves bi-pinnately compound (3- to 4-pinnate in some genera and sometimes only pinnate in *Gleditsia*); flowers regular or nearly so. 2
1 Leaves simple or pinnately compound; flowers papilionaceous. 8

2 Flowers in globose heads or cylindrical spikes; leaflets mostly less than ½ inch (12 mm) long. 3
2 Flowers in racemes; stamens 10 or less, their filaments free. 6

3 Twigs unarmed; legume compressed, dehiscent; flowers in heads, whitish . . . . **LYSILOMA** (*Lysiloma*)
3 Twigs armed with spines; filaments free; legume terete or compressed, indehiscent . . 4

4 Flowers in cylindric spikes, greenish- white, stamens 10; petioles minutely glandular at apex and tipped with small spinescent rachis. . . . . **MESQUITE** (*Prosopis*)
4 Flowers in heads or spikes, yellow or white, stamens more than 10; petioles not spiny or glandular at apex. 5

5 Plant armed with prickles . . . . **CATCLAW ACACIA** (*Senegalia*)
5 Plant armed with stipular spines . . . . **SWEET ACACIA** (*Vachellia*)

6 Flowers not showy, greenish-white, polygamous or dioecious; leaflets over ½ inch (12 mm) long; eastern trees, but widely planted . . . . **HONEYLOCUST** (*Gleditsia*)
6 Flowers large and showy, yellow, perfect; leaflets less than ½ inch (12 mm) long; shrubs or small trees 7

7 Stamens red and long-exserted; legume broad and flat; introduced and naturalized . . . . **BIRD-OF-PARADISE SHRUB** (*Erythrostemon*)
7 Stamens inserted; native. . . . . **PALOVERDE** (*Parkinsonia*)

8 Leaves simple. 9
8 Leaves pinnately compound 10

9 Leaves heart-shaped; flowers in fascicles, red, with free stamens; spineless. . . . . **REDBUD** (*Cercis*)
9 Leaves oblong or absent; flowers in racemes, blue, stamens with 9 united anthers; twigs reduced to spines. . . . . **SMOKETHORN** (*Psorothamnus*)

10 Leaflets 1 to 4 inches (2.5 to 10 cm) long 11
10 Leaflets less than 1 inch (2.5 cm) long. 13

11 Leaves 3-foliate; flowers red. . . . . **CORALBEAN** (*Erythrina*)
11 Leaves many-foliate; flowers white, yellow, or blue 12

12 Twigs with stipular spines; legumes compressed, not constricted. . . . **LOCUST** (*Robinia*)
12 Twigs unarmed; legumes terete, constricted. . . . . **MESCALBEAN** (*Dermatophyllum*)

13 Leaves with glandular dots . . . . **KIDNEYWOOD** (*Eysenhardtia*)
13 Leaves without glandular dots. . . . . **TESOTA** (*Olneya*)

California Redbud *Cercis occidentalis* Torr.

ALSO CALLED **western redbud, Arizona redbud, Judas-tree**

SYNONYM *Cercis arizonica* Rose, *Cercis occidentalis* var. *orbiculata* (Greene) Tidestr.

DESCRIPTION Large shrub or small tree to 15 feet tall, with many spreading branches and rounded crown.

TWIGS reddish brown when young, becoming dark gray the second year, without hairs.

LEAVES rounded, 2 to 4 inches (5 to 10 cm) long and broad, heart-shaped at base and rounded at tip, edges without teeth, thick, dark green above and paler beneath, usually hairless.

FLOWERS appearing on old twigs before the leaves, numerous and very showy, several in a cluster, pealike, ½ to ¾ inch (12 to 20 mm) long, purplish pink, in March and April.

PODS 2 to 3 inches (5 to 7.5 cm) long, flat, thin, brown or purplish.

BARK smooth, gray, becoming slightly fissured.

WOOD yellowish brown with thin whitish sapwood.

HABITAT Rare and local, restricted to a few scattered localities in canyons and mountains in upper desert and woodland zones, 4,000 to 6,000 feet elevation; not in New Mexico.

NOTE California redbud is common at Grand Canyon. It is cultivated as an ornamental for the spectacular flowers which cover the branches in early spring. The astringent bark of has been used as a remedy for diarrhea and dysentery.

ETYMOLOGY *Cercis,* the Greek name for a European species; traditionally the tree on which Judas hanged himself.

CALIFORNIA REDBUD

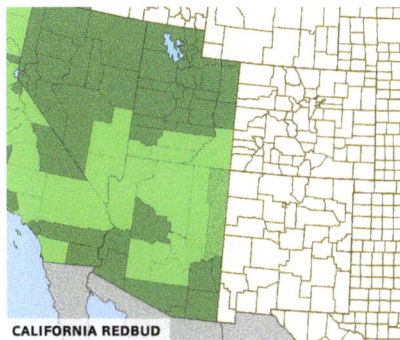

CALIFORNIA REDBUD

CALIFORNIA REDBUD

CALIFORNIA REDBUD - FLOWERS

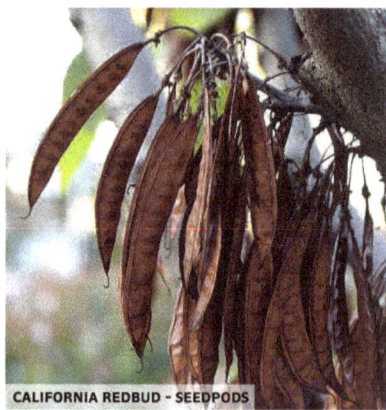

CALIFORNIA REDBUD - SEEDPODS

Mescalbean *Dermatophyllum secundiflorum* (Ortega) Gandhi & Reveal

ALSO CALLED frijolito, mescalbean sophora, coralbean, goat-bean

SYNONYM *Broussonetia secundiflora* Ortega, *Sophora secundiflora* (Ortega) Lag.

DESCRIPTION Evergreen shrub usually 3 to 5 feet tall or rarely small tree to 15 feet tall and 4 inches in trunk diameter, much branched.

TWIGS densely white-hairy when young.

LEAVES pinnately compound, 3 to 6 inches (7.5 to 15 cm) long. Leaflets 7 to 13, elliptical, ¾ to 2 inches (2 to 5 cm) long, rounded or slightly notched at tip, leathery, shiny dark green above, paler and hairless or nearly so beneath.

FLOWERS in dense dropping clusters 2 to 3 inches (5 to 7.5 cm) long, large and showy, ¾ to 1 inch long, pealike, purple, fragrant, in April.

POD 1 to 4 inches (2.5 to 10 cm) or more in length, ½ inch (12 mm) in diameter, finely white-hairy, hard and thick-walled, narrowed between the seeds, with 1 to 4 or more rounded scarlet seeds ½ inch (12 mm) long.

WOOD hard, very heavy, orange with red streaks and with thick yellow sapwood.

HABITAT Local in canyons and limestone cliffs of foothills and mountains, oak woodland zone, 4,500 to 6,500 feet elevation; Arizona populations considered adventive from further east.

NOTE This attractive shrub with beautiful blue flowers and conspicuous red seeds is reported to be cultivated in the southern States. The seeds are poisonous; a single seed, if eaten, may be sufficient to kill an adult. Also, the foliage is thought to be toxic to livestock.

ETYMOLOGY *Broussonetia,* the former name, honors *Pierre Marie August Broussonet* (1761-1807), Professor of Botany at Montpellier, France.

MESCALBEAN

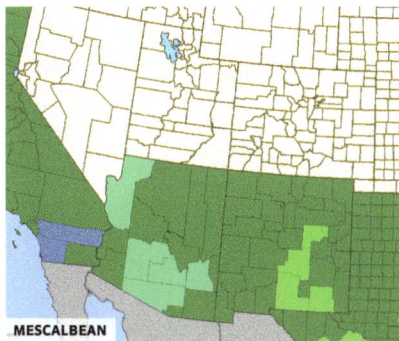

MESCALBEAN

Southwestern Coralbean *Erythrina flabelliformis* Kearney

ALSO CALLED western coralbean, chilicote, Indian-bean

DESCRIPTION Spiny shrub or sometimes a small tree to 15 feet tall and 10 inches (25 cm) in diameter at base, leafless except in summer.

TWIGS thick, brittle, with many scattered, single or paired, short, hooked spines.

LEAVES pinnately compound, with spiny leafstalk and 3 leaflets, broadly triangular with rounded angles or fan-shaped, 1½ to 3 inches (3.5 to 7.5 cm) long and 2 to 4 inches (5 to 10 cm) broad, gray-green, produced in summer.

FLOWERS several in a cluster, large and showy, 1½ to 2 inches (3.5 to 5 cm) long and only ¼ inches (6 mm) broad, bright red or scarlet, in spring and sometimes also late summer.

POD large, 6 to 10 inches (15 to 25 cm) long and ½ to ¾ inch (12 to 20 mm) broad, thick-walled, with several large, bright red seeds about ½ inch (12 mm) long.

BARK light brown with longitudinal white lines.

WOOD very brittle.

HABITAT Fairly common on warm, dry rocky slopes, washes, and canyons of foothills and mountains in upper desert, desert grassland, and oak woodland zones, 3,000 to 5,000 feet elevation, mostly in the Mexican border region.

NOTE Usually shrubby with branches freezing back in severe winters. The scarlet seeds are poisonous. They are used in making novelties and necklaces, and formerly by Native Americans as a charm.

ETYMOLOGY From Greek, *erythros*, red; an allusion to the color of the flower.

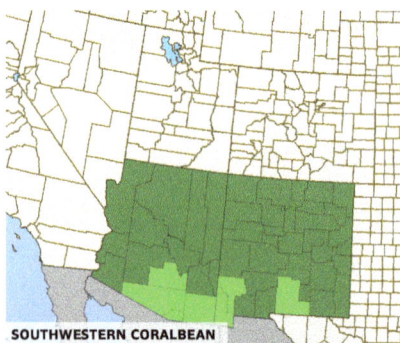

SOUTHWESTERN CORALBEAN

SOUTHWESTERN CORALBEAN

SOUTHWESTERN CORALBEAN

SOUTHWESTERN CORALBEAN

Bird-Of-Paradise Shrub *Erythrostemon gilliesii* (Wall. ex Hook.) Klotzsch

SYNONYM *Caesalpinia gilliesii* (Wall. ex Hook.) Wall. ex D. Dietr., *Poinciana gilliesii* Wall. ex Hook.

DESCRIPTION Introduced shrub or rarely a small tree, 3 to 12 feet tall (1 to 4 m); plants with a foul odor, stems with gland-tipped hairs.

LEAVES evergreen, glabrous, bipinnately compound,

FLOWERS yellow with orange marks; petals 5, with 10 pairs of showy, long red stamens, borne in racemes 4 to 8 inches (10 to 20 cm) long . April or May to August and September.

FRUIT an oblong, flat pod, often twisted when mature; dotted with a covering of short, red, gland-tipped hairs.

HABITAT Native to South America, introduced as an ornamental in the Southwest, where sometimes escaping to disturbed areas, roadsides, and gravelly areas; elevations mostly less than 3,000 feet.

NOTE The seeds and green seed pods are toxic if eaten, inducing severe vomiting and other abdominal distress.

BIRD-OF-PARADISE SHRUB

BIRD-OF-PARADISE SHRUB

Kidneywood　*Eysenhardtia polystachya* (Ortega) Sarg.

SYNONYM *Eysenhardtia orthocarpa* (A. Gray) S. Watson

DESCRIPTION Shrub or sometimes a small tree to 20 feet in height, with a short trunk 6 inches (15 cm) or more in diameter and many slender branches.

TWIGS gray or brown, hairy when young.

LEAVES pinnately compound, 2 to 5 inches (5 to 12.5 cm) long, resinous and gland-dotted, with disagreeable odor. Leaflets mostly 10 to 23 pairs with the last unpaired, narrowly oblong, ⅜ to ¾ inch (9 to 20 mm) long, rounded or slightly notched at tip, thick, pale gray green, finely hairy, with conspicuous brown dots beneath.

FLOWERS numerous, crowded in clusters 3 to 4 inches (7.5 to 10 cm) long, stalkless and pointed downward, nearly ½ inch (12 mm) long, white, with minute brown dots, from May to August.

PODS many, ½ inch (12 mm) long, flat and slightly curved, brown, hanging down, 1-seeded.

BARK thin, scaly and peeling off, light gray.

Wood heavy, hard, light reddish brown with thin yellow sapwood.

HABITAT Common on dry hillsides and rocky canyons of mountains, upper desert grassland, and lower oak woodland, 3,500 to 5,000 feet elevation.

NOTE The resinous foliage is readily browsed by livestock and wildlife. Water in which the heartwood has been soaked has a peculiar fluorescence and formerly was used for kidney and bladder diseases under the name *lignum nephriticum.*

KIDNEYWOOD

KIDNEYWOOD

Honey-Locust *Gleditsia triacanthos* L.

DESCRIPTION Deciduous tree to over 100 feet tall m tall, crown broad and open. Trunk single, reaching 2-4 feet diameter; trunk and larger branches often armed with thorns, these stout, sharp, and often branched.

LEAVES alternate, pinnately and bipinnately compound; blade to about 6 inches (15 cm) long, terminal segment and terminal leaflets absent; bipinnate leaves with 2 to 8 pairs of segments, each with 5 to 10 pairs of leaflets; pinnate leaves with 10 to 14 pairs of leaflets. Leaflets elliptic, ½ to 1½ inches (12 to 35 mm) long, margins entire or minutely toothed.

FLOWERS male or female on same tree, small, fragrant, greenish yellow. April to June.

POD large, flat, often curved, twisted, or contorted, 8 to 10 inches (20 to 45 cm) long and ¾ to 1¾ inches (2 to 4 cm) wide, reddish brown or blackish.

HABITAT Open woods and riparian areas, to about 1,300 feet elevation. Native to much of the eastern and central USA, considered adventive in our region.

NOTE Recognized by the pinnate and bipinnate leaves; large, often branched thorns, and long pods. Despite its name, the honey locust is not an important honey plant; the common name derives from the sweet taste of the pod's pulp, which was used for food and medicine by Native Americans. Thornless forms are often planted as ornamentals.

ETYMOLOGY Named for *Gottlieb Gleditsch* (d. 1786), Director of the Botanical Gardens, Berlin.

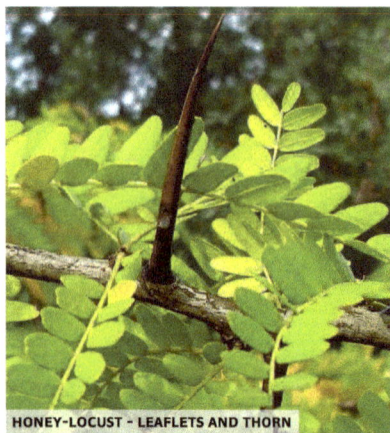

HONEY-LOCUST - LEAFLETS AND THORN

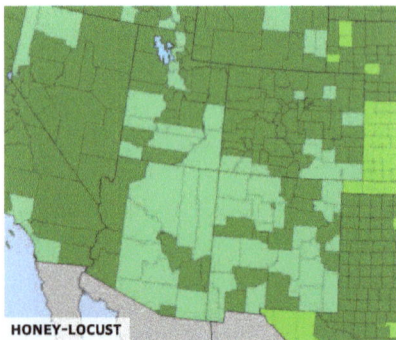

HONEY-LOCUST

Littleleaf Lysiloma　*Lysiloma watsonii* Rose

SYNONYM *Lysiloma microphylla* Benth., *Lysiloma thornberi* Britton & Rose

DESCRIPTION Shrub to 10 feet tall, dying back in severe winters; in cultivation a small tree 10 to 15 feet tall and 5 inches (12.5 cm) in trunk diameter, with spreading crown of dense feathery foliage; in Mexico a tree to 40 feet in height.

LEAVES numerous, bipinnately compound, 4 to 6 inches (10 to 15 cm) or more in length, with 4 to 9 pairs of primary divisions (pinnae), each with 25 to 33 pairs of feathery, oblong leaflets about ¼ inch (6 mm) long, nearly hairless or slightly hairy.

FLOWERS crowded into balls ⅝ inch (15 mm) in diameter, whitish, in May and June.

POD 4 to 9 inches (10 to 23 cm) long, broad and flat. Bark fissured and scaly, light brownish gray.

WOOD hard, brittle, dark brown.

HABITAT Very rare and local on rocky hillsides in upper desert and desert grassland, 2,800 to 4,000 feet elevation, known from several localities in southern Arizona; not in New Mexico.

NOTE The plants of this Mexican species at its northern limit in southern Arizona may be relics of wider distribution in the past. They have large bases and are killed back by cold winters. This handsome shrub is planted at the Boyce Thompson Southwestern Arboretum at Superior, Arizona; it is suitable as an ornamental in warmer areas, although subject to occasional winter kill.

LITTLELEAF LYSILOMA

LITTLELEAF LYSILOMA

Tesota *Olneya tesota* A. Gray

ALSO CALLED desert ironwood, palo fierro, palo de hierro

DESCRIPTION Spiny evergreen tree to 30 feet in height, with a short trunk to 3 feet in diameter with symmetrical, widely spreading crown.

TWIGS green or gray and densely hairy when young, becoming light brown, with sharp spines about ¼ inch (6 mm) long in pairs at nodes.

LEAVES in dense clusters, pinnately compound, 1 to 2¼ inches (2.5 to 5.5 cm) long, generally evergreen. Leaflets 2 to 10 pairs, oblong or obovate, ⅜ to ¾ inches (9 to 20 mm) long, usually rounded at ends, thick, blue-green, finely hairy with pressed hairs.

FLOWERS in short clusters, pealike, ½ inch (12 mm) long, purplish, in May and June.

POD 2 to 2½ inches (5 to 6 cm) long, brown, glandular hairy, thick-walled, with few seeds.

BARK thin, smoothish, gray, on large trunks becoming much fissured and shreddy.

WOOD very hard and heavy, dark brown with thin yellow sapwood. In weight, it is perhaps second only to leadwood (southern Florida), of all trees native in the USA. It is easily polished and is used for novelties, such as bowls and small boxes. Many trees have been cut for fuel.

HABITAT Common and characteristic desert tree of sandy washes in foothills at lower, warmer elevations from slightly above sea level to 2,500 feet; not in New Mexico.

NOTE Limited to warm areas, and regarded as an indicator of sites suitable for citrus orchards. The seeds, which are produced in quantities, are edible when roasted, and have been used for food by Native Americans and local residents.

ETYMOLOGY Named for *Stephan Olney* (1812-1878), an American botanist.

TESOTA

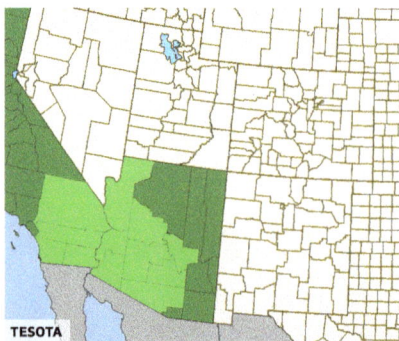

TESOTA

Parkinsonia PALOVERDE

Large shrubs or small trees, attaining a height of about 25 feet (8 m); young bark smooth, green; rachis of the pinnae short, terete; flowers showy, yellow, in corymbose fascicles; pods more or less torulose (swollen and constricted at intervals).

Paloverdes are common and characteristic plants of the lower and drier parts of the Southwest, and are a glorious sight when in full flower. They are leafless in the dry season but conspicuous because of their greenish bark. The wood is soft and brittle and burns very quickly, giving off an unpleasant odor and leaving few coals. The pods are fairly palatable but are not much eaten by livestock, except during prolonged droughts. It is reported that Native Americans ate the seeds after grinding them into meal. The flowers produce a good honey.

Named for *John Parkinson* (1567-1650), apothecary of London and author of several important botanical books.

1   Leaflets alternate; rachis of the pinnae flattened, 4 inches (10 cm) long or longer; inflorescence an elongate raceme up to 8 inches (20 cm) long .. JERUSALEM-THORN
    . . . . (*Parkinsonia aculeata*)
1   Leaflets opposite; rachis of the pinnae terete, not more than 2 inches (5 cm) long; inflorescence a short raceme or a corymb. 2

2   Young bark yellowish green; leaves appearing simply pinnate, the common rachis nearly obsolete; leaflets of each pinna commonly 4 to 8 pairs, these very small, not more than ⅛ inch (3 mm) long; flowers pale yellow, the odd petal often whitish; pod turgid, tipped with a flat triangular or sword-shaped beak . . . . . . **YELLOW PALOVERDE**
    . . . . (*Parkinsonia microphylla*)
2   Young bark bluish green; leaves evidently bipinnate, the common rachis short but apparent; leaflets of each pinna 1 to 3 pairs, ⅛ inch to ⅜ inch (4 to 8 mm) long; flowers bright yellow, all of the petals alike in color; pod flat, with a short triangular beak (or almost beakless). . . . . **BLUE PALOVERDE** (*Parkinsonia florida*)

Jerusalem-Thorn *Parkinsonia aculeata* L.

ALSO CALLED Mexican paloverde, horsebean, retama

DESCRIPTION Spiny, small or medium-sized tree to 40 feet tall and 1 foot in trunk diameter, with smooth, yellow-green bark, branches, and twigs, very open crown of spreading branches and drooping twigs and "streamers," appearing as if leafless most of year.

TWIGS slightly zigzag, slender, with paired short spines at nodes bordering a third, larger, brownish spine 1 to 1½ inch (2.5 to 3.5 cm) long at end of the very short leaf axis.

LEAVES bipinnately compound but appearing as if pinnate, consisting of the short, spinetipped axis and 1 to 3 pairs of wiry but flattened, narrow, evergreen "streamers" (pinnae) 8 to 20 inches (20 to 50 cm) long. Leaflets 25 to 30 pairs on a "streamer," narrowly oblong, ¼ inch or less in length, light green, soon falling.

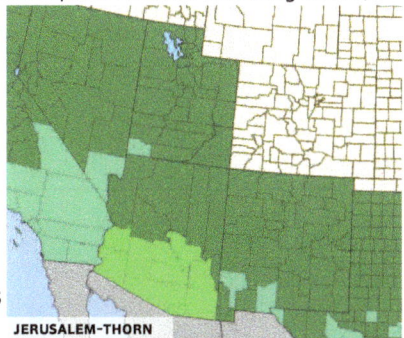

JERUSALEM-THORN

Flowers in loose clusters 3 to 8 inches (7.5 to 20 cm) long, showy, about ¾ inch (12 mm) across, golden yellow except that upper petal is red-spotted and turns red in withering.

POD 2 to 4 inches (5 to 10 cm) long, less than ½ inch (12 mm) in diameter, long-pointed, narrowed between the few seeds.

BARK smooth, thin, yellow green, at base of larger trunks becoming scaly and brown.

WOOD hard and heavy.

HABITAT Rare and very local as a native tree in Arizona, in foothills, canyons, and valleys of desert and desert grassland zones, 3,000 to 4,500 feet elevation; also extensively planted as an ornamental tree and escaping. Not native in New Mexico but reported as an escape from cultivation in southern part. Naturalized from Florida to Texas and California and in tropical parts of the world.

NOTE Jerusalem-thorn is a popular, fast-growing, attractive tree widely planted as an ornamental in warmer parts of the Southwest and in tropical countries.

ETYMOLOGY The word "Jerusalem" in this and other plant names (such as Jerusalem-artichoke), does not refer to the city in Palestine but is a corruption of the Spanish word *girasol*, meaning "turning toward the sun."

JERUSALEM-THORN

Blue Paloverde *Parkinsonia florida* (Benth. ex A. Gray) S. Watson

SYNONYM *Cercidium floridum* Benth., *Cercidium torreyanum* (S. Wats.) Sarg., *Parkinsonia torreyana* (Benth.) S. Wats.

DESCRIPTION Spiny, small tree to 30 feet in height and 1½ feet in trunk diameter, with widely spreading, very open crown and with smooth, blue-green bark, branches, and twigs, leafless most of the year.

TWIGS slightly zigzag, bearing a spine ¼ inch (6 mm) or less in length at each node.

LEAVES few and scattered, bipinnately compound, 1 inch (2.5 cm) long, with 1 pair of primary divisions (pinnae), each bearing 2 or 3 pairs of oblong leaflets about ¼ inch (6 mm) long, pale bluegreen and hairless at maturity, appearing in spring but soon shedding.

FLOWERS numerous over the tree, 4 or 5 in a cluster 1 to 2 inches (2.5 to 5 cm) long, about ¾ inch (2 cm) across, bright yellow, the upper petal with a few red spots but not turning red in withering, in late March or April (sometimes also from August to October).

POD 3 to 4 inches (7.5 to 10 cm) long, flattened.

BARK smooth, thin, blue-green, at base of larger trunks becoming scaly and brown.

WOOD soft, heavy, light brown with light yellow sapwood.

HABITAT Common and characteristic tree along washes and valleys and sometimes on slopes, desert and desert grasslands, from slightly above sea level to 4,000 feet elevation; not in New Mexico.

NOTE Blue paloverde is very showy when in blossom, becoming a mass of bright yellow flowers.

BLUE PALOVERDE

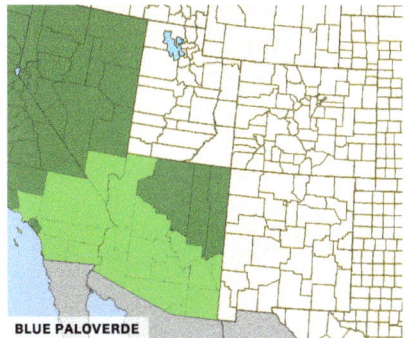

BLUE PALOVERDE

Yellow Paloverde *Parkinsonia microphylla* Torr.

ALSO CALLED **foothill paloverde, littleleaf paloverde, littleleaf horsebean**

SYNONYM *Cercidium microphyllum* (Torr.) Rose & Johnst.

DESCRIPTION Spiny small tree to 25 feet tall and 1 foot in trunk diameter, with widely spreading, much-branched open crown, and with smooth, yellow-green bark, branches, and twigs, leafless most of the year. Short lateral twigs stiff and ending in stout spines.

LEAVES few, bipinnately compound but apparently pinnately compound, ¾ to 1 inch (20 to 25 mm) long, consisting of a very short axis and 1 pair of primary divisions (pinnae, each bearing 4 to 7 pairs of very small elliptical leaflets ⅛ inch (3 mm) or less in length, yellow-green, slightly hairy, appearing in spring but soon shedding).

FLOWERS numerous and covering the tree, few in a cluster 1 inch long (2.5 cm), about ½ inch (12 mm) across, pale yellow, about ½ inch (12 mm) in diameter, the upper petal white or cream, in April and May.

POD 2 to 3 inches (5 to 7.5 cm) long, about ¼ inch (6 mm) in diameter, cylindric, 1- to 3-seeded, narrowed between the seeds.

BARK smooth, thin, yellow green, at base of larger trunks becoming slightly rough and gray.

WOOD hard, heavy, dark orange brown streaked with red, with thick light brown or yellow sapwood.

HABITAT Abundant and characteristic tree on plains, foothills, and mountains in Arizona deserts, 500 to 4,000 feet elevation; not in New Mexico.

NOTE Yellow paloverde, or "foothill paloverde," is a characteristic, slowly growing tree of Arizona deserts, associated especially with the saguaro. This species is commoner than blue paloverde and grows on the widespread drier sites, while blue paloverde occurs chiefly along drainages. Both species are leafless most of the year. The Spanish common name "palo verde," meaning "green tree," "green pole," or "green stick," refers to the distinctive smooth green branches and twigs, which manufacture food in the absence of leaves and with much less evaporation of water. The specific names yellow paloverde and blue paloverde refer to the color shades of the branches, twigs, and foliage. Yellow paloverde blossoms later in the spring than blue paloverde and has

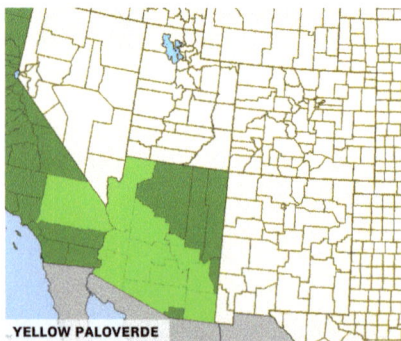

YELLOW PALOVERDE

smaller and paler yellow flowers. Seeds of paloverdes were ground and used as food by the Native Americans. The trees are grown as ornamentals.

YELLOW PALOVERDE

YELLOW PALOVERDE

YELLOW PALOVERDE

YELLOW PALOVERDE

Prosopis MESQUITE

Shrubs or small trees, usually armed with straight spines; leaves with 2 to 4 pinnae, and numerous narrow leaflets; flowers in cylindric spikes, small, greenish yellow, somewhat fragrant; pods indehiscent, compressed but somewhat turgid, or spirally coiled.

The flowers furnish the best of nectar for honey making. The leaves and pods are eaten by all kinds of grazing animals. The large roots and thickened bases of the stems furnish the best firewood of the region (only *Olneya* is harder). The legumes (seed pods) and seeds were collected by Native Americans, who ground them and formed the meal into a sort of bread.

Mesquite increases rapidly on overgrazed grassland; the hard seeds, which remain viable in the soil many years, are widely spread by livestock eating the pods. In sandy soils of southern New Mexico, dunes often form around shrubby mesquites, burying the plants except for a hemispherical mass of branching tips. *Prosopis* is the Greek name for the butter-bur, but the origin is obscure.

1 Pods tightly spirally coiled, ¾ to 1¾ inches (2 to 4 cm) long; leaflets 5 to 9 pairs; spines white, slender; herbage sparsely to copiously grayish pubescent. . . . . SCREWBEAN MESQUITE (*Prosopis pubescens*)
1 Pods not coiled, compressed, more or less constricted between the seeds, much more than 1¾ inches (4 cm) long; leaflets commonly many more than 9 pairs; spines, if any, yellowish, often stout 2

2 Foliage pubescent, leaflets less than ⅝ inch (15 mm) long . . . . VELVET MESQUITE . . . . (*Prosopis velutina*)
2 Foliage glabrous or nearly so, leaflets mostly more than ⅝ inch (15 mm) long . . . . HONEY MESQUITE (*Prosopis glandulosa*)

Honey Mesquite *Prosopis glandulosa* Torr.

SYNONYM *Prosopis juliflora* (Sw.) DC.

DESCRIPTION Spiny, spreading shrub branched from the base or a small to medium-sized tree 20 to 50 feet tall with a trunk 1 to 4 feet in diameter and with spreading crown of crooked branches.

TWIGS slightly zigzag, hairless (or in velvet mesquite densely short-hairy or velvety), with stout yellowish spines ¼ to 1 inch (6 to 25 mm) long in pairs at nodes, the

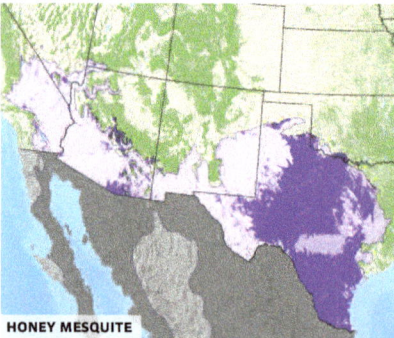

HONEY MESQUITE

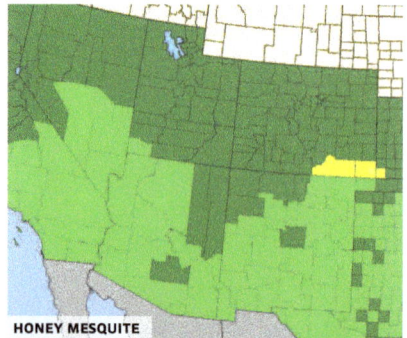

HONEY MESQUITE

nodes afterwards bearing short knotlike spurs ¼ to ½ inch (6 to 12 mm) in diameter.

LEAVES bipinnately compound, 3 to 8 inches (7.5 to 20 cm) long, with 1 or 2 pairs of primary divisions (pinnae), each with 9 to 22 pairs of yellow-green leaflets, usually very narrow and hairless (oblong and finely hairy in velvet mesquite).

FLOWERS in dense narrow clusters 2 to 3 inches (5 to 7.5 cm) long, yellow, fragrant, from April to August.

POD 3 to 8 inches (7.5 to 20 cm) long, hairless (finely hairy in velvet mesquite), not splitting open, with sweetish pulp.

BARK rough and thick, separating into long narrow strips, dark brown.

WOOD hard and heavy, reddish brown with thin yellow sapwood.

HABITAT Common to abundant on sandy plains and sandhills, and along stream valleys and washes, in short-grass, desert, desert-grassland, and (infrequently) lower oak-woodland zones, from slightly above sea level to 5,500 feet elevation.

HONEY MESQUITE

Screwbean Mesquite *Prosopis pubescens* Benth.

ALSO CALLED **Fremont screwbean, screwbean mesquite, tornillo**

SYNONYM *Prosopis odorata* Torr. & Frem., *Strombocarpa odorata* (Torr. & Frem.) Torr., *Strombocarpa pubescens* (Benth.) A. Gray

DESCRIPTION Spiny shrub or small tree to 20 feet tall and 1 foot in trunk diameter, or sometimes larger; less common than our other mesquites.

TWIGS usually gray-hairy when young, with white or pale gray stout spines ⅛ to ¾ inch (3 to 20 mm) long, paired at nodes and united to base of the leaf-stalks.

LEAVES bipinnately compound, 1 to 3 inches (2.5 to 7.5 cm) long, with 1 or sometimes 2 pairs of primary divisions (pinnae), each with 5 to 8 pairs of oblong, slightly hairy leaflets about ⅜ inch (9 mm) long.

FLOWERS in usually dense clusters 2 to 3 inches (5 to 7.5 cm) long, from May to August.

POD 1 to 1½ inches (2.5 to 3.5 cm) long, less than ¼ inch (6 mm) in diameter, tightly coiled like a large screw, brown, not splitting open, with sweetish pulp.

BARK separating in long fibrous strips, light brown or reddish.

WOOD very hard, heavy, light brown; used for firewood, fence posts, and locally for tool handles.

HABITAT Common along watercourses and in valleys in desert zone, from slightly above sea level to 5,500 feet elevation.

NOTE Easily recognized by the screwlike pods, and often called *tornillo*, Spanish for screw. The sweet pods can be chewed and eaten, and were made into meal and cakes by Native Americans.

SCREWBEAN MESQUITE

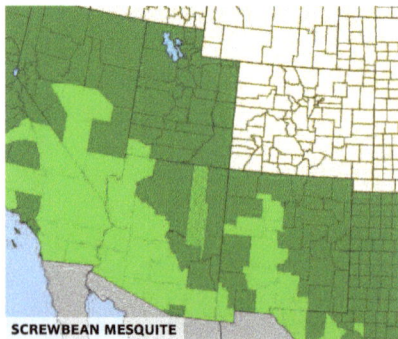

SCREWBEAN MESQUITE

SCREWBEAN MESQUITE – PODS

SCREWBEAN MESQUITE

Velvet Mesquite *Prosopis velutina* Woot.

SYNONYM *Prosopis juliflora* var. *velutina* (Woot.) Sarg.
Usually a medium-sized tree.,

DESCRIPTION Deciduous spiny tree or large shrub, 9–15 m tall, with a round crown of twisted branches. Trunk low-branching, to 2 feet diameter.

LEAVES bipinnately compound, grayish green; usually some leaves forked into 2 pairs (sometimes 1 or 3) of segments to 6 inches (15 cm) long; leaflets 28–48 per segment, elliptic-oblong, each ⅜ to ½ inch (9 to 12 mm) long, almost touching one another, minutely velvety hairy; with paired, straight, stipular spines ½ to ¾ inch (1 to 2 cm) long at nodes.

FLOWERS pale greenish cream, in spikes, very similar to those of honey mesquite; from March to May.

POD 4 to 8 inches (10 to 20 cm) long, tan, mottled with red or black

BARK at first smooth and greenish; later dark brown to blackish, shredding into shaggy strips and flakes.

WOOD Sapwood yellow; heartwood heavy, reddish-brown, hard and slow burning.

HABITAT Arid grasslands, outwash fans, along watercourses and roadsides; from low-elevations to about 5,500 feet.

TIP Velvet mesquite has finely short-hairy to velvety twigs, leaflets, and pods; it also has the smallest (only ¼ to ½ inch [6 to 12 mm] long) and more closely spaced leaflets (about ⅛ inch [3 mm] apart) than leaflets in our other species (in honey mesquite, the leaflets are larger, more widely separated, and hairless).

NOTE This plant has long been a mainstay for Southwestern Native Americans; a meal called *pinole,* made from the long sweet pods, prepared in the form of cakes and in other ways, was a staple food.

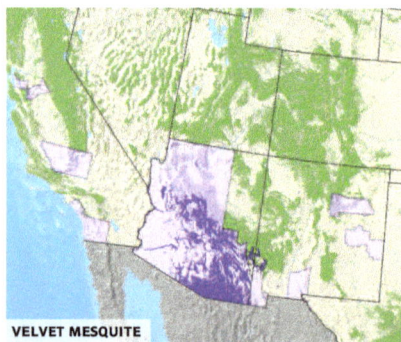

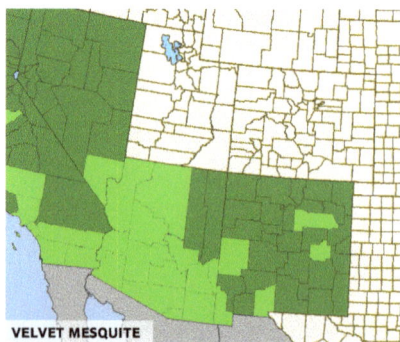

VELVET MESQUITE

VELVET MESQUITE

VELVET MESQUITE

VELVET MESQUITE

Smokethorn *Psorothamnus spinosus* (A. Gray) Barneby

ALSO CALLED smoketree, indigobush, smokethorn dalea

SYNONYM *Dalea spinosa* A. Gray

DESCRIPTION Small, compact, spiny tree or large shrub to 20 feet tall, with a short, branching trunk up to 1½ feet in diameter and with compact mass of smoky gray or silvery branches and spiny twigs, leafless most of the year.

TWIGS gray or silvery with dense, pressed hairs, with brown gland dots, ending in sharp spines.

LEAVES few, reverse lance-shaped, ¾ to 1 inch (2 to 2.5 cm) long, gland-dotted, remaining only a few weeks and falling before flowering.

FLOWERS few in a cluster, pealike, ½ inch (12 mm) long, dark purple, with brown gland dots, in April and May.

PODS egg-shaped, ⅜ inch (9 mm) long, gland dotted, with 1 or 2 seeds.

BARK of trunk thin, fissured and scaly, dark gray brown.

WOOD soft, lightweight, dark brown with whitish sapwood.

HABITAT Washes in the desert at low, frost-free elevations from slightly above sea level to 1,000 feet (rarely 1,500 feet); not in New Mexico.

NOTE Smokethorn, or "smoketree," derives its common names from the smoky gray color of the compact, leafless branches and twigs, as seen from a distance.

SMOKETHORN

SMOKETHORN

SMOKETHORN

SMOKETHORN

New Mexico Locust *Robinia neomexicana* A. Gray

ALSO CALLED Southwestern locust

DESCRIPTION Spiny shrub or small tree up to 25 feet tall and 8 inches in trunk diameter.

TWIGS with brownish, glandular hairs when young, with brown or reddish, stout spines ¼ to ½ inch (6 to 12 mm) long in pairs at nodes.

LEAVES pinnately compound, 6 to 12 inches (15 to 30 cm) long. Leaflets 15 to 21, elliptical, ½ to 1½ inches (12 to 35 mm) long, bristle-tipped, thin, pale blue green, hairy when young but becoming nearly hairless.

FLOWERS many in large clusters, large and showy, about ¾ inch (2 cm) long, pealike, purplish pink, fragrant, from May to August.

POD 2½ to 4½ inches (6 to 12.5 cm) long, flat, thin, reddish brown, bristly hairy and usually glandular; the inner pulp toxic.

BARK furrowed and scaly, thin, light brown or gray.

WOOD very hard, heavy, yellow streaked with brown.

HABITAT Common to abundant in canyons and mountains, often forming almost pure thickets on north-facing slopes, and also associated with Gambel oak, ponderosa pine forest, and upper pinyon-juniper woodland; 4,000 to 8,500 feet elevation.

NOTE Occasionally propagated as an ornamental. Because of rapid growth and the tendencies to form thickets and to sprout from the horizontal roots, the plants are of special value in reducing erosion.

ETYMOLOGY Named for *Jean Robin* (d. 1629), herbalist to Henri IV of France.

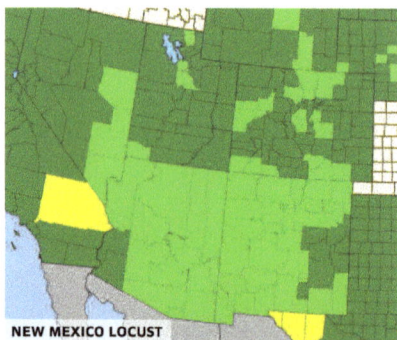

NEW MEXICO LOCUST

NEW MEXICO LOCUST

NEW MEXICO LOCUST

NEW MEXICO LOCUST

Catclaw Acacia *Senegalia greggii* (A. Gray) Britton & Rose

ALSO CALLED catclaw, devilsclaw, una de gato

SYNONYM *Acacia greggii* A. Gray

DESCRIPTION Spiny, large spreading shrub or occasionally a small tree to 23 feet tall and 8 inches in trunk diameter, much-branched, with broad crown.

TWIGS light reddish brown or purplish, angled, finely hairy, with scattered stout curved spines ¼ inch or less in length.

LEAVES bipinnately compound, 1 to 3 inches (2.5 to 7.5 cm) long, with 2 or 3 pairs of primary divisions (pinnae), each with 4 to 6 pairs of oblong, usually hairy leaflets ¼ inch or less in length.

FLOWERS in dense clusters 1 ½ to 2 inches (3.5 to 5 cm) long and ½ inch (12 mm) in diameter, pale yellow, fragrant, from April to October.

POD 2½ to 5 inches (6 to 12.5 cm) long and ½ inch (12 mm) broad, flat, pale brown, often twisted and narrowed between the shiny brown seeds, remaining attached in winter.

BARK thin, fissured into narrow scales, gray.

WOOD hard and heavy, the heartwood reddish brown and sapwood yellow.

HABITAT Often common to abundant, forming thickets along washes, slopes, and rocky canyons in desert and desert grassland, from slightly above sea level to about 5,000 feet elevation.

NOTE Catclaw acacia is to be avoided because of the sharp stout spines or prickles like a cat's claws, which tear clothing and flesh. The strong, heavy wood with contrasting heartwood and sapwood is used for souvenirs, locally for tool handles and similar objects, and also for fuel. The flowers are a good source of honey. Native Americans made meal from the seeds.

ETYMOLOGY The former genus name, *Acacia,* is the Greek name for the tree (derived from *akis,* a sharp point).

CATCLAW ACACIA

CATCLAW ACACIA

CATCLAW ACACIA

Sweet Acacia *Vachellia farnesiana* (L.) Wight & Arn.

ALSO CALLED **huisache**

SYNONYM *Acacia farnesiana* (L.) Willd.

DESCRIPTION Spiny shrub or small tree to 20 feet tall and 1 foot in diameter at base, with widely spreading crown.

TWIGS with white, straight, slender spines ⅛ to ½ inch (3 to 12 mm) long or larger in pairs at nodes.

LEAVES bipinnately compound, 2 to 4 inches (5 to 10 cm) long, with usually 3 to 6 pairs of primary divisions (pinnae), each with 10 to 25 pairs of narrow leaflets ⅛ to ¼ inch (3 to 6 mm) long, often slightly hairy.

FLOWERS in balls ⅜ inch (9 mm) in diameter, golden yellow, very fragrant, from April to November.

POD 2 to 2½ inches (5 to 6 cm) long and ¼ to ⅜ inch (6 to 9 mm) in diameter, not flattened, hard.

BARK thin, ridged and scaly, reddish brown.

WOOD hard and heavy, reddish brown with thin pale sapwood.

HABITAT Very rare in desert grassland and lower oak woodland, 3,500 to 5,000 feet elevation, known from only several localities in southern Arizona; not in New Mexico. Widely planted and naturalized in southern United States from Florida to Texas and California. Also in Mexico, Central America, and South America.

NOTE Cultivated as an ornamental because of the highly sweet-scented, fragrant flowers. In southern Europe extensively grown for the flowers, known as *cassie flowers,* from which perfume is made.

ADDITIONAL SPECIES **Mescat acacia**, or "white-thorn" (*Vachellia constricta* (Benth.) Seigler & Ebinger, **synonym:** *Acacia constricta* Benth.), a closely related species, is a common, smaller, branching and spreading shrub 2 to 10 feet (0.6 to 3 m) tall, with longer white spines often 1 to 1½ inches (2.5 to 3.5 cm) long and

SWEET ACACIA

with pods flattened and narrowed between the seeds. It is widely distributed in the desert and desert grassland of southern New Mexico and southern and central Arizona.

SWEET ACACIA

FAGACEAE *Beech Family*

Quercus OAK

Trees or shrubs; leaves alternate, simple, petioled, the blades entire, toothed, or lobed, persistent or deciduous; flowers monoecious; staminate flowers in drooping catkins; pistillate flowers solitary or clustered; fruit a nut (acorn), partly enveloped by an involucre (cup), maturing in 1 or 2 seasons. Hybrids between the various oaks of our region, having intermediate characteristics bertween the parent species, are quite common. *Quercus* is the Latin name.

The shrubby or scrub oaks, such as *Quercus turbinella,* are the main component of the chaparral on exposed mountainsides in southern and central Arizona and New Mexico. At higher elevations, especially in northern portions of the region, the deciduous white oaks (principally *Quercus gambelii*), are abundant. Acorns are an important food for wildlife, and Native Americans roasted the nuts, then mixing them with meat or fat. Although not important as a commercial lumber source in the Southwest, some oaks are used locally for firewood and fence posts.

NOTE Key adapted from the *Flora of North America* website (*www.efloras.org*).

1 Mature bark smooth or deeply furrowed, not scaly or papery; cup scales flattened; leaf blade if lobed then with awned teeth, if entire then often with bristle at tip **2**
1 Mature bark scaly or papery, rarely deeply furrowed; cup scales thickened; leaf blade, if lobed, without awned teeth, if unlobed without bristle at the tip **3**

2 Leaf underside glabrous except for tufts of tomentum in vein axils or at base; cup covering ⅓ or more of nut; Arizona and New Mexico. . . **EMORY OAK** (*Quercus emoryi*)
2 Leaf underside tawny or white-tomentose; cup covering ⅓ of nut (or less); southern Arizona, southwestern New Mexico . . . . **SILVERLEAF OAK** (*Quercus hypoleucoides*)

3 Acorn requiring 2 seasons to mature; cup scales embedded in tawny or glandular hairs, only scale tips visible; plants evergreen, leaves leathery, glaucous (waxy) on leaf underside **4**
3 Acorn requiring 1 growing season to mature; cup scales not embedded in tawny or glandular hairs; plants evergreen or deciduous. **5**

4 Twigs stiff, mostly ¹⁄₁₆ to ⅛ inch (1.5 to 3 mm) diameter, branching at 65–90° angles; cup scales joined into concentric rings; acorn oblong to fusiform; Arizona, southwestern New Mexico . . . . **PALMER OAK** (*Quercus palmeri*)
4 Twigs flexible, stouter, mostly ⅛ to ³⁄₁₆ inch (3 to 4 mm) diameter, branching at less than 60° angles; cup scales not in noticeable concentric rings; acorn ovoid; Arizona, rare in southwestern New Mexico . . . . **CANYON LIVE OAK** (*Quercus chrysolepis*)

5 Leaf blade regularly or irregularly lobed, at least some sinuses extending more than ⅓ distance from margin to midrib; Arizona and New Mexico . . . . **GAMBEL OAK** . . . . (*Quercus gambelii*)
5 Leaf blade unlobed or sometimes shallowly lobed or wavy-lobed; if lobed, sinuses less than ⅓ distance to midrib **6**

6 Leaf blade with 10 to 20 ±parallel secondary veins on each side, these ±straight and not branching, margins regularly toothed, the teeth never spine-tipped; eastern New Mexico . . . . **CHINQUAPIN OAK** (*Quercus muehlenbergii*)
6 Leaf blade with 2 to 12 irregular secondary veins on each side, these curved, crooked, and/or branching, or obscure, not noticeably parallel; margins entire or irregularly toothed or spinose, if toothed then the teeth not evenly spaced. **7**

7 Leaves and twigs hairless at maturity, leaf blade glaucous and blue-green to gray-green **8**

7 Mature leaves pubescent, hairy or glandular on undersurface (visible with 10× lens), twigs variously hairy or sometimes only sparsely so, leaf blade greenish to grayish or blue-green **9**

8 Leaf margins with spine-tipped teeth; leaf underside papillose (with many small bumps), visible under strong magnification; southern and central Arizona . . . . **AJO MOUNTAIN OAK** (*Quercus ajoensis*)

8 Leaf blade unlobed or wavy-lobed, margins entire or obscurely toothed, not sharply toothed or spine-tipped; southern Arizona, southwestern New Mexico . . . . **MEXICAN BLUE OAK** (*Quercus oblongifolia*)

9 Mature leaf underside yellowish green, without stellate or erect bundles of hairs, or with spreading hairs only along midevin and near base of blade; southern Arizona, southwestern New Mexico . . . . **TOUMEY'S OAK** (*Quercus toumeyi*)

9 Mature leaf underside with stellate or erect bundles of hairs, these distributed ± evenly (the hairs sometimes tiny or obscured by glandular hairs or wax) **10**

10 Leaf blade usually convexly cupped, leaf underside with prominent raised veins; secondary veins often impressed on upper surface of leaf **11**

10 Leaf blade underside even, without prominent raised veins; secondary veins not strongly impressed on upper surface of leaf **12**

11 Acorn on stalk 1¼ to 2½ inch (3 to 6 cm) long, usually 3 to several acorns per stalk; leaf blade broadly obovate to orbiculate; Arizona, western New Mexico . . . . **NETLEAF OAK** (*Quercus rugosa*)

11 Acorn nearly stalkless or on axillary peduncle less than 1¼ inch (3 cm) long , usually 1 to 2 (sometimes 3) acorns per stalk; leaf blade elliptic to lanceolate, ovate, or obovate; Arizona and New Mexico . . . . **ARIZONA WHITE OAK** (*Quercus arizonica*)

12 Leaf blade undersurface with obvious or minute appressed-stellate hairs, sometimes appearing glabrous or glaucous or yellowish glandular to naked eye, not felty or velvety to touch. **13**

12 Leaf blade undersurface with erect fasciculate or (nonappressed) erect or semi-erect stellate hairs, these often felty or velvety to touch **14**

13 Acorn on stalk ⅛ to 1¼ inch (3 to 30 mm) long; upper leaf surface smooth and waxy; acorn ¾ inch (2 cm) long; Arizona and New Mexico . . . . **SHRUB LIVE OAK** . . . . (*Quercus turbinella*)

13 Acorn nearly stalkless or on short stalk to ¼ inch (6 mm) long; upper leaf surface rough and "sandpapery"; acorn to ⅜ to ⅝ inch (1 to 1.5 cm) long; southern Arizona, New Mexico . . . . **SANDPAPER OAK** (*Quercus pungens*)

14 At least some leaf margins undulate, with 2 to 3 rounded teeth on each side; plants usually low, colony-forming shrubs, often on stabilized sand dunes; eastern New Mexico. . . . . **HAVARD'S OAK** (*Quercus havardii*)

14 Leaf margins not undulate (flat or undulate in *Q. mohriana*), entire or with sharp teeth; plants trees or shrubs, on various substrates **15**

15 Bud dark red-brown, glabrous or occasionally finely hairy on outer scales; acorn nearly stalkless or on stalk ⅜ to ⅝ inch (1 to 1.5 cm) long; plants of limestone and calcareous substrates; eastern New Mexico . . . . **MOHR'S OAK** (*Quercus mohriana*)

15 Bud yellowish because of stellate hairs (at least on outer scales); acorn often on stalk to ¾ to 1¼ inch (2 to 3 cm) long; plants of igneous substrates; Arizona and New Mexico . . . . **GRAY OAK** (*Quercus grisea*)

Arizona White Oak   *Quercus arizonica* Sarg.

ALSO CALLED **Arizona oak**

DESCRIPTION Medium-sized evergreen tree 30 to 60 feet or more in height and 2 to 3 feet in trunk diameter, with irregular spreading crown of stout branches.

LEAVES obovate or oblong, 1 to 3 inches (2.5 to 7.5 cm) long, short-pointed or rounded at tip, heart-shaped or rounded at base, edges slightly wavy-lobed and toothed toward tip, thick and stiff, above dull blue-green and nearly hairless and with veins sunken, beneath paler and densely hairy and with prominent raised veins, shedding gradually in spring as the new leaves unfold.

ACORN ¾ to 1 inches (2 to 2.5 cm) long, with shallow cup.

BARK fissured into thick plates, light gray or whitish.

WOOD hard, heavy, dark brown.

HABITAT Common and characteristic tree of the oak woodland, in foothills, mountains, and canyons, 5,000 to 7,600 feet elevation.

NOTE Arizona white oak is probably the largest of the southwestern oaks, reaching its greatest size in canyons and other moist sites. This species is closely related to *Quercus grisea* and is reported to form hybrids with it and other oak species.

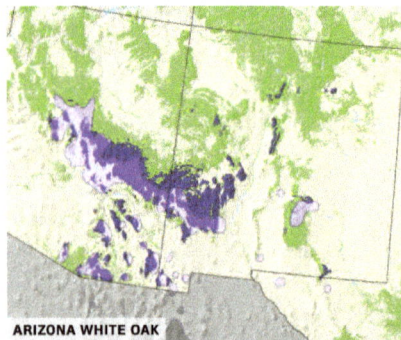

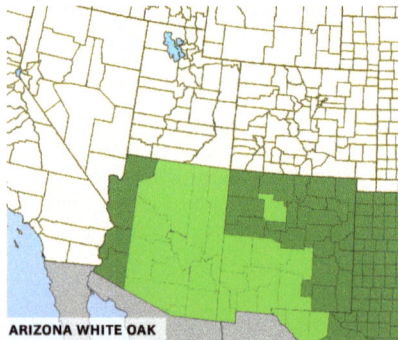

ARIZONA WHITE OAK

ARIZONA WHITE OAK

ARIZONA WHITE OAK

ARIZONA WHITE OAK

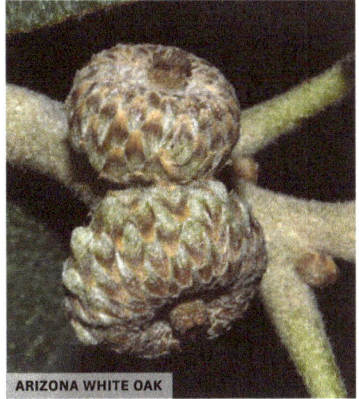

ARIZONA WHITE OAK

Canyon Live Oak *Quercus chrysolepis* Liebm.

DESCRIPTION A wide-spreading, evergreen oak, rarely more than 60 feet high, usually a tall shrub in our region (more often tree-like in California, where a common oak).
TWIGS flexible, diverging at 60° or less from the branch, golden velvety-hairy when young, becoming gray-hairy or hairless when older.
LEAVES simple, thin, somewhat stiff, ovate or oblong, ¾ to 2¾ inch (2 to 7 cm) long; flat or somewhat rolled; usually sharply spine-tipped; margins entire or with low, spine-tipped teeth; upper surface yellowish green, lustrous, smooth; lower surface waxy blue-white, sometimes with gold-colored glandular hairs.
ACORN red-brown, oval or elliptic in outline, ⅝ to 1¼ inch (1.5 to 3 cm) long; stalk absent or to only ¹⁄₃₂ inch (1 mm) long; the cup often yellowish.
BARK light gray to gray-brown, thin and smooth when young; older trunks furrowed, with small scales.
WOOD hard, heavy, light brown.
HABITAT Wooded canyons and slopes; low to mid-elevations.
ETYMOLOGY *chrysolepis,* golden-scale, refers to the yellowish acorn cups.

CANYON LIVE OAK

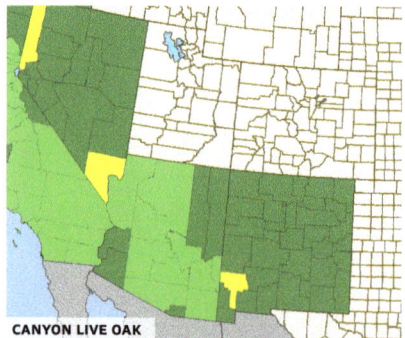

CANYON LIVE OAK

CANYON LIVE OAK

CANYON LIVE OAK

CANYON LIVE OAK

Emory Oak *Quercus emoryi* Torr.

ALSO CALLED black oak, blackjack oak, bellota

DESCRIPTION Medium-sized evergreen tree to 50 feet in height and 2½ feet in trunk diameter.

LEAVES broadly lance-shaped, 1 to 2½ inches (2.5 to 6 cm) long, with a short spiny point and a few short spiny teeth, thick, stiff, leathery, flat, shiny yellowish green on both sides, nearly hairless, shedding gradually in spring as the new leaves unfold.

ACORN ½ to ¾ inch long, rounded, one-third or more enclosed by the cup, sweetish and edible.

BARK thick, divided into thick plates, black.

WOOD hard and heavy, dark brown with thick lighter sapwood.

HABITAT The most abundant and most characteristic tree of the oak woodland in the Mexican border region, in foothills, mountains, and canyons and sometimes also in canyons at upper edge of desert and desert grassland, 4,000 to 7,000 feet (rarely 8,000 feet) elevation.

NOTE Found in moist places, especially in canyons, Emory oak may form fairly dense forests. It is one of the most important sources of firewood in southern Arizona. The acorns, known in Spanish as *bellotas,* are nearly free from tannin with its bitter taste and are gathered and eaten locally by Native Americans and Mexicans.

ETYMOLOGY Named for *Lt. Col. W. H. Emory* (1811-87), leader of a military expedition in the Southwest in 1846-47 (during the Mexican War), who first collected specimens of this species at Pigeon Creek (Las Palomas), within present-day Sierra County, New Mexico.

EMORY OAK

EMORY OAK

EMORY OAK

EMORY OAK

EMORY OAK

EMORY OAK

Gambel Oak *Quercus gambelii* Nutt.

ALSO CALLED Utah white oak, Rocky Mountain white oak

SYNONYM *Quercus gunnisonii* (Torr.) Rydb., *Quercus leptophylla* Rydb., *Quercus novomexicana* (A. DC.) Rydb., *Quercus submollis* Rydb., *Quercus utahensis* (A. DC.) Rydb.

DESCRIPTION Small to medium-sized tree 20 to 70 feet tall with trunk 2 to 4 feet in diameter and with rounded crown, or a shrub as low as 6 feet in height and growing in thickets.

LEAVES oblong, 2 to 7 inches (5 to 17 cm) long, deeply 7- to 11-lobed halfway or more to middle, edges without teeth, dark green above, light green and soft-hairy beneath, varying greatly in size, lobing, and hairiness, turning yellow and reddish in autumn before shedding.

ACORN ⅜ to ¾ inch (9 to 20 mm) long, broad and rounded, with deep cup.

BARK rough, thick, deeply furrowed or scaly, gray.

WOOD hard, heavy, light brown; the wood is used for fence posts and fuel.

HABITAT Common and widespread in mountains and plateaus in ponderosa pine forest, 5,000 to 8,000 feet elevation.

NOTE Gambel oak is easily recognized by the deeply lobed leaves, which are larger than those of other southwestern oaks and which are shed in autumn. This is the only common tree oak in the northern parts of New Mexico and Arizona.

GAMBEL OAK

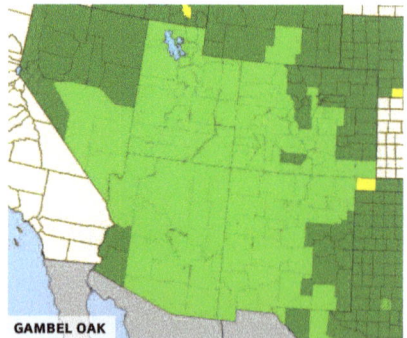

GAMBEL OAK

ETYMOLOGY This familiar southwestern oak (and the equally familiar Gambel quail) honor *William Gambel,* a physician, ornithologist, and botanist of Philadelphia, who travelled through Santa Fe in 1841 on his way to California, and discovered nearly one hundred new plant species during the course of the expedition.

GAMBEL OAK

GAMBEL OAK

Gray Oak *Quercus grisea* Liebm.

DESCRIPTION Small, low scrubby evergreen tree or shrub, or in favorable locations a medium-sized tree to 65 feet in height.

LEAVES elliptic to ovate, ¾ to 2 inches (2 to 5 cm) long, blunt or short-pointed at tip, rounded or slightly heart-shaped at base, edges without teeth or with a few teeth toward tip, thin and firm, gray green or blue green, shiny and sparsely hairy above, beneath densely hairy.

ACORN ½ inch (12 mm) long, rounded, one-third enclosed by the deep cup.

BARK fissured into shaggy plates, light gray.

WOOD hard, heavy, brown.

HABITAT Dry rocky mountain slopes and foothills and in canyons, oak woodland and pinyon-juniper woodland, 5,000 to 7,000 feet elevation.

NOTE Gray oak, usually a low, scrub oak, is common in the oak woodland of New Mexico and infrequent westward in Arizona. As this species intergrades both with Arizona white oak and shrub live oak, some forms are not readily distinguished.

GRAY OAK

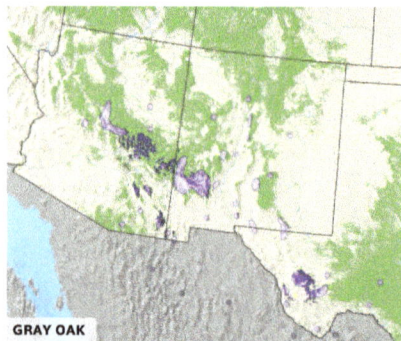

GRAY OAK

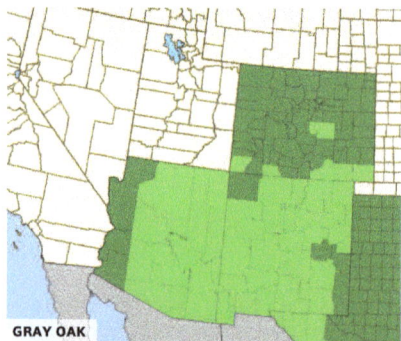

GRAY OAK

Havard's Oak   *Quercus havardii* Rydb.

ALSO CALLED **shinnery oak**

DESCRIPTION Colony-forming evergreen shrub with a normal height of 3 feet; rarely becoming a small tree.

TWIGS brown-hairy when young, becoming smooth with age.

LEAVES thick, ovate to elliptic, 2 to 4 inches (5 to 10 cm) long; margins usually with lobed with 2 or more rounded teeth on each side; upper surface shiny, light green; lower surface densely brown-hairy, the secondary veins quite prominent.

ACORN ovoid, ⅝ to 1¼ inch (1.5 to 3 cm) long, single or paired on short stalks less than ⅜ inch (9 mm) long.

BARK papery, light gray.

HABITAT pinyon-juniper and desert scrublands, usually in light, sandy soils; stabilized dunes; mid-elevations.

ETYMOLOGY Named for U.S. Army surgeon and botanist, *Valery Havard.*

NOTE Havard oak ranges from eastern New Mexico to the Texas Panhandle and western Oklahoma. As it often occurs on privately owned grazing lands, it is subject to herbicide spraying.

HAVARD'S OAK

HAVARD'S OAK

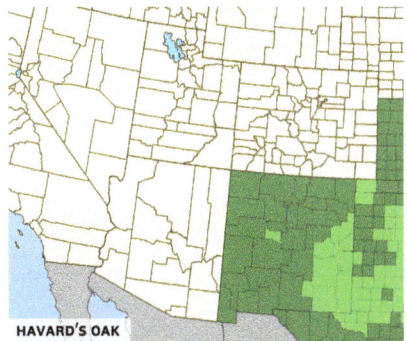

HAVARD'S OAK

Silverleaf Oak *Quercus hypoleucoides* A. Camus

ALSO CALLED whiteleaf oak

SYNONYM *Quercus hypoleuca* Engelm., not Miq.

DESCRIPTION Small to medium-sized evergreen tree to 30 feet to 65 feet in height and 1½ to 2½ feet in trunk diameter, with a rounded spreading crown, or sometimes a clump-forming shrub 6 feet tall.

LEAVES lance-shaped, 2 to 4 inches (5 to 10 cm) long and ½ to 1 inch (12 to 25 mm) wide, sharp-pointed, narrowed at base, edges rolled under and usually without teeth or with a few small spiny teeth, very thick and leathery, shiny yellow-green above, beneath densely white-woolly.

ACORN ½ to ⅝ inch (12 to 15 mm) long, pointed, one-third enclosed in a thick cup hairy inside, maturing in 2 years.

BARK deeply furrowed into ridges and plates, blackish.

WOOD hard, heavy, dark brown.

HABITAT Mountain slopes and canyons of oak woodland, 5,000 to 7,000 feet elevation, most common in Mexican border region.

NOTE Silverleaf oak, with its leaves so distinct from those of most oaks, is suitable for cultivation as an ornamental tree.

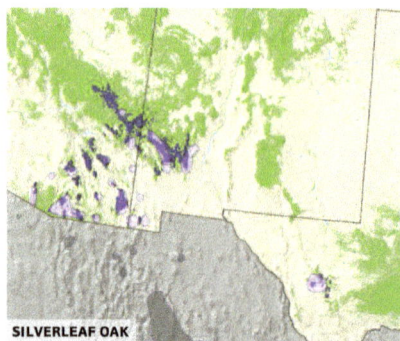

SILVERLEAF OAK

SILVERLEAF OAK

SILVERLEAF OAK

SILVERLEAF OAK – LEAF UNDERSIDE

SILVERLEAF OAK

Mohr's Oak *Quercus mohriana* Buckley ex Rydb.

ALSO CALLED Mohr's shinoak

DESCRIPTION Evergreen, thicket-forming shrub, occasionally a small tree to 20 feet high.

BARK thick, gray, with rough scaly ridges.

TWIGS densely grayish-brown hairy.

LEAVES thick and leathery, oblong to elliptical, 1⅛ to 3 inches (3 to 7.5 cm) long, rounded or acute at tip; shiny dark green above, grayish hairy on underside, beneath with prominently raised secondary veins; margin usually entire and wavy, sometimes with a few teeth.

ACORN light brown, ellipsoid to ovoid, ⅜ to ⅝ inch (8 to 15 mm) long, solitary or paired, nearly stalkless or the stalk to about ½ inch (12 mm) long; cup enclosing half the acorn.

HABITAT calcareous soils; low- to mid-elevations.

NOTE 'Shin oak' and 'shinnery' refer to the dense thickets, scarcely knee-high, of dwarf evergreen oaks of this and related species on uplands of western Texas and other areas of the Southwest.

ETYMOLOGY Named for botanist *Charles Mohr*, who wrote about the flora of Alabama.

MOHR'S OAK

MOHR'S OAK

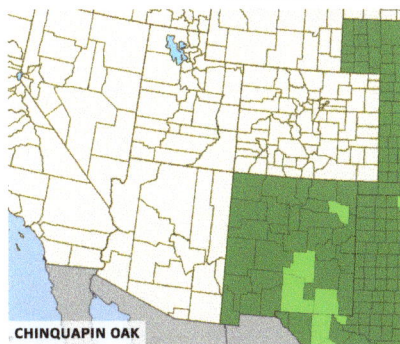

Chinquapin Oak   *Quercus muehlenbergii* Engelm.

SYNONYM *Quercus muhlenbergii* Engelm.

DESCRIPTION Small tree to 25 feet in height (in eastern United States a large tree).

LEAVES oblong or broadly lance-shaped, 3 to 6 inches (7.5 to 15 cm) long, short or long-pointed, usually rounded at base, edges wavy with coarse, slightly curved teeth, dark or yellowish green above, paler and finely hairy beneath.

ACORN ½ to ¾ inch (12 to 20 mm) long, rounded, half enclosed by the deep cup.

BARK thin, fissured, and flaky, light gray.

WOOD heavy and hard.

HABITAT Very rare and local in mountain canyons of pinyon-juniper woodland, about 7,000 feet elevation, in mountains bordering the plains in eastern New Mexico; not in Arizona. Also widely distributed in eastern United States from northwestern Florida north to Vermont and southern Ontario, west to Wisconsin and Iowa, and south to Texas.

NOTE Chinquapin oak is too rare in New Mexico to be important but is noteworthy because of its unusual distribution. These isolated Rocky Mountain localities are some distance westward from the westernmost main range of this species in western Oklahoma and central Texas.

ACORN

CHINQUAPIN OAK

CHINQUAPIN OAK

Mexican Blue Oak _Quercus oblongifolia_ Torr.

DESCRIPTION Small evergreen tree to 25 feet tall, with trunk up to 1½ feet in diameter and with spreading, rounded crown, or a shrub at higher elevations.

LEAVES oblong, 1 to 2 inches (2.5 to 5 cm) long, rounded at both ends or heart-shaped at base, edges without teeth (rarely toothed), thin and firm, at maturity without hairs, blue-green and covered with a bloom above, paler beneath.

ACORN to ½ to ¾ inch (12 to 20 mm) long, rounded, one-third enclosed by the cup.

BARK fissured into small, squarish plates, gray.

WOOD hard, very heavy, brittle, dark brown.

HABITAT Common and characteristic small tree of oak woodland in foothills and mountains and sometimes in canyons at upper edge of desert and desert grassland, 4,500 to 6,000 feet elevation, mostly in Mexican border region.

NOTE Mexican blue oak, which forms open woodlands of small spreading trees along the Mexican border, is readily distinguished by the small, blue-green, oblong, toothless and hairless leaves.

MEXICAN BLUE OAK

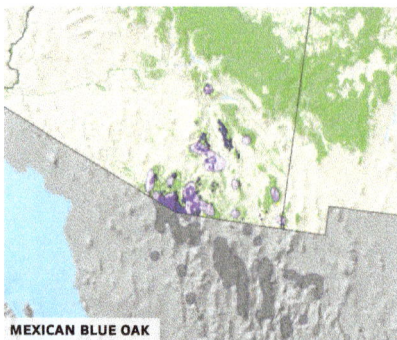

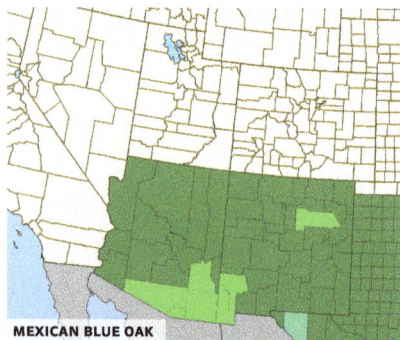

MEXICAN BLUE OAK

MEXICAN BLUE OAK

MEXICAN BLUE OAK

Palmer Oak *Quercus palmeri* Engelm.

ALSO CALLED canyon live oak

SYNONYM *Quercus chrysolepis* var. *palmeri* (Engelm.) Sarg.

DESCRIPTION Evergreen shrub or small tree, usually 6 to 25 feet tall and up to 6 inches in trunk diameter, with a dense, bushy, broad crown of many stiff branches.

LEAVES elliptic to ovate, ¾ to 2 inches (2 to 5 cm) long, short-pointed, edges crisp and spine-toothed, stiff and very leathery, shiny yellow-green above, beneath slightly yellowish and resinous hairy but becoming whitish.

ACORN ⅝ to 1½ inches (15 to 35 mm) long, maturing in 2 years; the cup large, and loosely fitting the acorn, densely hairy inside and out with a coat of golden hairs.

BARK fissured into narrow scales and flakes, gray or brown.

WOOD hard, heavy, light brown.

HABITAT Canyons and mountainsides, often forming thickets, elevation 3,500 to 7,000 feet.

NOTE Palmer oak has spiny evergreen leaves resembling those of holly and is attractive in winter; the stiff twigs make Palmer oak thickets very difficult to penetrate.

NOTE Populations in eastern Arizona and southwestern New Mexico are often intermediate between this species and the closely related *Q. chrysolepis.*

ETYMOLOGY Honors *Edward Palmer* (1831-1911), American botanical collector who collected numerous specimens in Mexico and southwestern United States during many years of exploration.

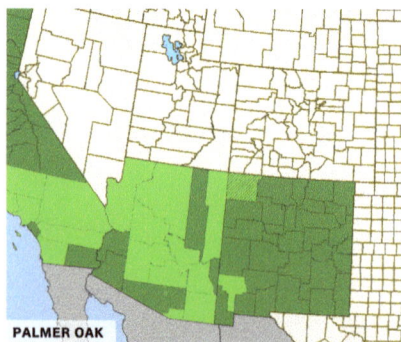

PALMER OAK

PALMER OAK

PALMER OAK

PALMER OAK - ACORN

Sandpaper Oak *Quercus pungens* Liebm.

ALSO CALLED pungent oak

DESCRIPTION Evergreen shrub usually 3 to 6 feet high, rarely a small tree.

TWIGS densely woolly when young, the older twigs becoming nearly hairless within about two years; purplish gray to gray.

LEAVES leathery, elliptic to oblong, 1⅝ to 2 inches (1.5-5 cm) long, undersurface densely woolly with prominent veins, upper surface yellowish-green; persisting about 1 year; margins strongly wavy, with 2 to 5 lobes per side, each sinus reaching about a quarter to more than half-way to the midvein.

ACORNS light brown, broadly ovoid to nearly cylindric, to about ½ inch (12 mm) long, solitary or in pairs; inner surface of cap, densely appressed hairy.

BARK light gray or brown, flaky when older.

HABITAT Often found on limestone soils, in chaparral, oak, and juniper communities, sometimes also into low-elevation desert vegetation; from 3,500 to 5,500 feet.

NOTE Can be confused with *Quercus palmeri,* but differs by having less round and more deeply-lobed leaves.

SANDPAPER OAK - ACORNS

SANDPAPER OAK

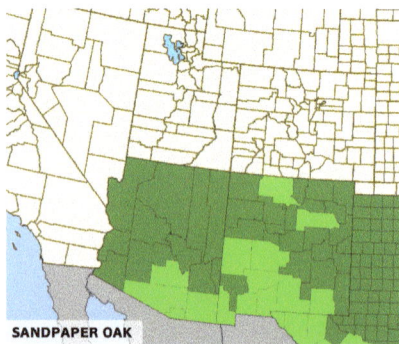

SANDPAPER OAK

Netleaf Oak *Quercus rugosa* Née

SYNONYM *Quercus diversicolor* Trel., *Quercus reticulata* Humb. & Bonpl.

DESCRIPTION Small to medium-sized evergreen tree to 40 feet tall, with broad rounded crown, or a shrub to only 6 feet in height.

LEAVES broadly obovate, varying in form and size, 1 to 4 inches (2.5 to 10 cm) long, rounded at tip and slightly heart-shaped at base, edges with several small spiny teeth especially toward tip, thick and stiff, above dark green and slightly hairy and with veins sunken, beneath paler and yellow hairy and with a network of raised veins.

FLOWERS of this and other oaks are male and female on the same tree, the male in narrow hanging clusters, the female 1 or few in leaf axils, in spring.

ACORNS 2 or 3 together on long stalk 1 to 4 inches (2.5 to 10 cm) long, ½ inch (12 mm) long, oblong, one-fourth enclosed by the shallow cup.

BARK fissured and flaky, gray.

WOOD hard, light brown.

HABITAT Uncommon in mountains and canyons, oak woodland zone, 4,000 to 8,000 feet elevation.

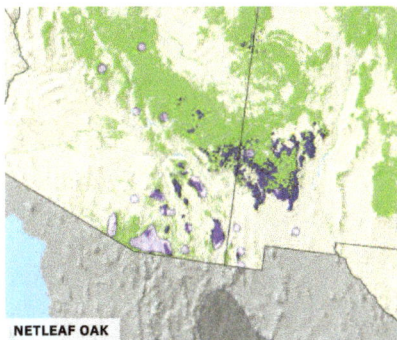

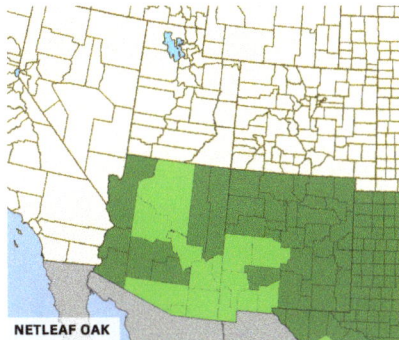

NETLEAF OAK - LEAF UNDERSIDE

NETLEAF OAK

NETLEAF OAK

Toumey's Oak *Quercus toumeyi* Sarg.

DESCRIPTION Evergreen shrub 3 to 6 feet tall or small tree to 33 feet in height and 8 inches (20 cm) in trunk diameter.

LEAVES numerous and crowded, very small, elliptic or oval, ½ to ¾ inch (12 to 20 mm) long, sharp-pointed, edges without teeth or sometimes with a few short teeth, shiny yellow-green above, slightly hairy beneath, shedding in spring as new leaves appear.

ACORN ½ to ⅝ inch (12 to15 mm) long, with shallow cup.

BARK thin, scaly or flaky, dark brown.

WOOD light brown.

HABITAT Local and restricted on hillsides and mountains in oak woodland, 4,000 to 7,000 feet elevation, Mexican border region. Shrubby Toumey's oaks may be seen in Texas Canyon along the highway east of Benson in Cochise County, Arizona.

TOUMEY'S OAK

NOTE Toumey's oak was discovered in the Mule Mountains of Cochise County, Arizona, in 1899 by *James W. Toumey* (1865-1932), then botanist at the University of Arizona and afterwards dean of the Yale University School of Forestry.

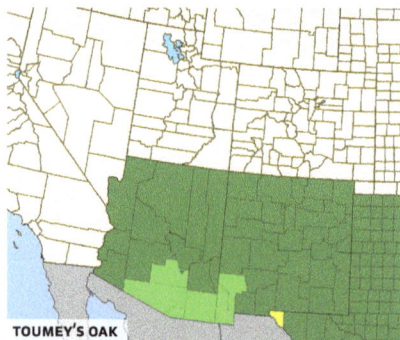

TOUMEY'S OAK

Shrub Live Oak　*Quercus turbinella* Greene

ALSO CALLED **scrub oak, California scrub oak, turbinella oak**

SYNONYM *Quercus dumosa* Nutt. var. *turbinella* (Greene) Jepson, *Quercus subturbinella* Trel.

DESCRIPTION Evergreen, much-branched shrub usually less than 8 feet tall, or rarely a small tree to 15 feet in height with an open, widely spreading crown.

LEAVES small, elliptic or oblong, ½ to 1¼ inch (12 to 30 mm) long, short-pointed, edges with many small spinelike teeth, thick and stiff, upper surface blue-green with a bloom and nearly hairless, beneath yellowish green and finely hairy.

ACORN ¾ inch (2 cm) long, narrow and pointed, with shallow cup.

BARK fissured and scaly, gray.

WOOD hard, brittle, brown with tan or yellowish sapwood.

HABITAT Abundant as the characteristic species of the chaparral in Arizona, forming dense thickets and covering hillsides, and also in oak woodland, pinyon-juniper woodland, lower ponderosa pine forest, and sometimes upper edge of desert, 4,500 to 8,000 feet elevation.

NOTE Though generally a shrub in the evergreen shrubby or chaparral vegetation, this species rarely becomes a small tree at Grand Canyon and elsewhere. Shrub live oak and other southwestern oaks provide browse for goats, sheep, and cattle.

SHRUB LIVE OAK

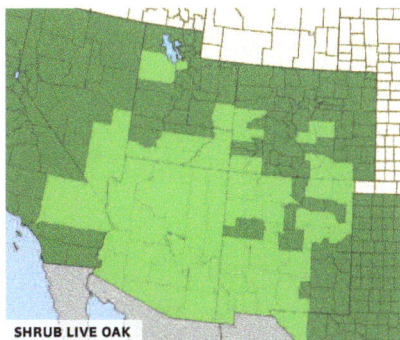

SHRUB LIVE OAK

ADDITIONAL SPECIES Ajo Mountain Oak (*Quercus ajoensis* C.H. Muller, synonym *Quercus turbinella* subsp. *ajoensis*).

Evergreen shrubs or low trees, 6 to 10 feet (2 to 3 m) high; **bark** gray, scaly or furrowed; **young twigs** pubescent. **Leaves** ovate, leathery, with toothed margins, the teeth spine-tipped, blue green on the upperside, sometimes slightly hairy on underside. **Acorns** solitary or paired, the cup shallow, enclosing only the base of the nut. Known from rocky slopes in southern Arizona, from 1,500-5,000 feet elevation. Distinguishing features are the ovate or oblong and leathery leaves to 1½ inches (3.5 cm) long, with 4-6 teeth on each side; the lower leaf surface is conspicuously glaucous with fine beads of bluish wax.

The species name derives from means of or from Ajo, where the type location is found in the Ajo Mountains near Organ Pipe Cactus National Monument.

Perhaps best considered as merely a variety or subspecies of *Quercus turbinella* as its characters intergrade with *Q. turbinella*.

AJO MOUNTAIN OAK

AJO MOUNTAIN OAK

SHRUB LIVE OAK

JUGLANDACEAE *Walnut Family*

Juglans WALNUT

A tree, or occasionally shrubby, with strong-scented, pinnately compound leaves; leaflets commonly 9 to 13, large, lanceolate or ovate-lanceolate, acuminate, serrate; staminate flowers in long drooping catkins; pistillate flowers solitary, or very few in a cluster; fruit a large, usually nearly globular, hard-shelled nut enclosed in a finally dry husk; cotyledons 2-lobed. *Juglans,* the Latin name, is from *jovis,* of Jupiter, and *glans,* an acorn.

1   Leaflets 9-13 (rarely to 19), oblong-lanceolate; nut 1 to 1½ inches (2.5 to 3.5 cm) long.
.... ARIZONA WALNUT (*Juglans major*)
1   Leaflets 17-23, narrow-lanceolate; nut ½ to ¾ inch (12 to 20 mm) long...............
.... LITTLE WALNUT (*Juglans microcarpa*)

Arizona Walnut *Juglans major* (Torr.) Heller

ALSO CALLED Arizona black walnut, nogal
SYNONYM *Juglans rupestris* var. *major* Torr.
DESCRIPTION Small to medium-sized tree 30 to 50 feet tall and 1 to 2 feet (rarely 4 feet) in trunk diameter, with rounded crown of widely spreading branches.
TWIGS reddish brown, densely hairy when young, becoming ashy gray.
LEAVES pinnately compound, 7 to 14 inches (18 to 36 cm) long, with characteristic strong walnut odor. Leaflets usually 9 to 13 (rarely to 19), lance-shaped or broadly lance-shaped, 2 to 4 inches (5 to 10 cm) long and ¼ to 1½ inches (6 to 35 mm) wide, long-pointed, often slightly curved, edges coarsely saw-toothed, thin, scurfy-hairy when young but becoming hairless or nearly so, yellow green.
FRUIT almost spherical, 1 to 1½ inches (2.5 to 3.5 cm) in diameter, with thin, densely

ARIZONA WALNUT

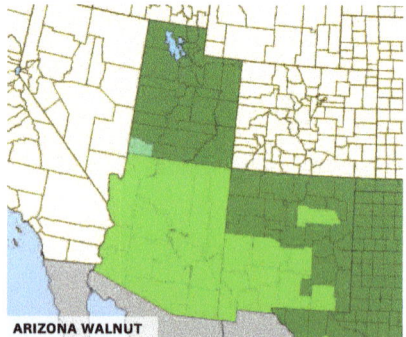

ARIZONA WALNUT

hairy, brown husk, a thick hard shell, and a small edible kernel.

BARK on small trunks smoothish and fissured, on large trunks thick, deeply furrowed and ridged, grayish brown.

WOOD hard, heavy, chocolate brown with thick whitish sapwood.

HABITAT Scattered along streams and canyons, mostly in mountains, in upper part of desert, desert grassland, and oak woodland zones, often with Fremont cottonwood, Arizona sycamore, Arizona alder, and willows, 3,500 to 7,000 feet elevation.

NOTE The small, thick-shelled walnuts, known in Spanish as *nogales,* are gathered and eaten by local residents. The trees are planted for shade. Enlarged burls and stumps at the base of some tree trunks are prized for the beautiful patterns of their wood and are manufactured into table tops and veneer. The wood is suitable for the same uses as black walnut, such as furniture, cabinets, and gunstocks, and has been cut in small quantities for cabinetmaking. However, the supply is limited, scattered, and not easily accessible, and the trees generally are small. The wood is durable and used locally for fence posts.

ARIZONA WALNUT

ARIZONA WALNUT · WINTER TWIG AND BUDS

Little Walnut *Juglans microcarpa* Berland.

ALSO CALLED **Texas black walnut, nogal**

SYNONYM *Juglans rupestris* Engelm.

DESCRIPTION Large branching shrub or small tree 10 to 20 feet tall and ½ to 1½ feet in trunk diameter, usually branching at or near ground, with broad rounded crown.

TWIGS reddish brown, densely hairy when young, becoming ashy gray.

LEAVES pinnately compound, 8 to 13 inches (20 to 32 cm) long, with characteristic walnut odor. Leaflets usually 17 to 23 (sometimes as few as 13), narrowly lance-shaped, 2 to 3 inches (5 to 7.5 cm) long and about ½ inch (12 mm) wide, long-pointed, usually slightly curved, edges finely saw-toothed or almost without teeth, thin, becoming hairless or nearly so, yellow green.

FLOWERS of this and the next species greenish, male in narrow hanging clusters and female few together, in spring.

FRUIT spherical, ½ to ¾ inch (12 to 20 mm) in diameter, with a thin, hairy, brown husk, thick hard shell, and a small edible kernel.

WOOD hard, heavy, chocolate brown with thick whitish sapwood.

HABITAT Scattered along streams in plains and foothills desert zone, 3,000 to 4,000 feet elevation; not in Arizona.

NOTE This species has the smallest nuts of all the walnuts, suggesting marbles in size.

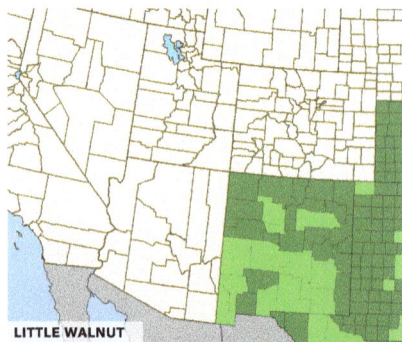

LITTLE WALNUT

LITTLE WALNUT

LITTLE WALNUT

KOEBERLINIACEAE *Crown-of-Thorns Family*

Koeberlinia is the sole genus in the family Koeberliniaceae; the genus was formerly placed in the Caper Family (Capparaceae).

Allthorn *Koeberlinia spinosa* Zucc.

ALSO CALLED crucifixion-thorn, crown-of-thorns, corona de Cristo, junco

DESCRIPTION Very spiny, much-branched, rounded spreading shrub 3 to 6 feet tall or in a variety (var. *tenuispina,* found in Arizona) also a small tree to 15 feet in height, with numerous, stout, widely forking, tangled, dark green branches and twigs, leafless most of the year.

TWIGS stout, ⅛ inch (3 mm) or more in diameter, dark or pale green, nearly hairless, ending in spines 1 to 2 inches (2.5 to 5 cm) long (2 to 4 inches (5 to 10 cm) long and more slender in var. *tenuispina*).

LEAVES scalelike, soon shedding.

FLOWERS few in small clusters on the twigs, less than ¼ inch (6 mm) long, greenish white, from March to June.

FRUIT a shiny black berry less than ¼ inch (6 mm) in diameter.

BARK dark green, becoming scaly and gray.

WOOD very hard, heavy, dark brown streaked with orange, with lighter sapwood.

HABITAT Clay or sandy plains, slopes and foothills, often forming dense thickets and commonly growing with creosotebush and tarbush ("blackbrush"), desert and desert grassland, 1,500 to 5,000 feet elevation.

NOTE Allthorn is one of the three shrubby species of "crucifixion-thorns" in the Southwest. Var. *tenuispina* of southwestern Arizona becomes a small tree and is distinguished by darker green twigs, longer spines 2 to 4 inches (5 to 10 cm) long, earlier flowering, and its occurrence at lower elevations and in more sandy soil.

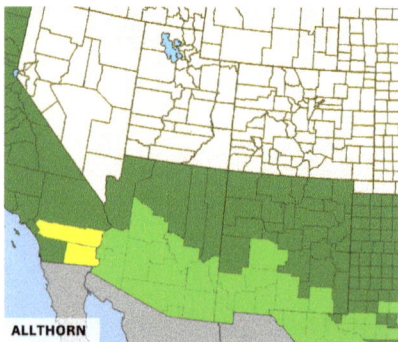

ALLTHORN

ALLTHORN

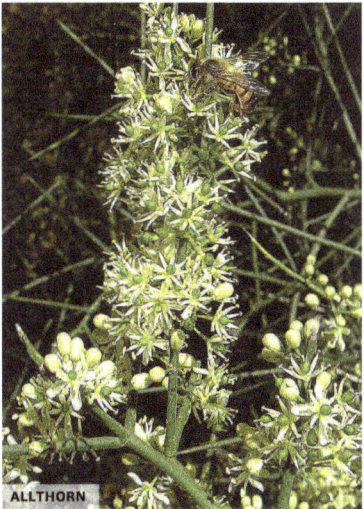

ALLTHORN

MALVACEAE *Mallow Family*

California Fremontia *Fremontodendron californicum* (Torr.) Cov.

ALSO CALLED flannelbush, mountain leatherwood, California slippery-elm

SYNONYM *Fremontia californica* Torr.

DESCRIPTION Evergreen shrub or sometimes a small tree 10 to 20 feet tall.

TWIGS stout, densely rusty-hairy when young, becoming light reddish brown and hairless.

LEAVES on short lateral twigs, broadly ovate or slightly heart-shaped, ½ to 1½ inches (12 to 35 mm) long and broad, usually 3-lobed with rounded lobes, thick, dull or dark green and sparsely hairy above, beneath rusty- and scurfy-hairy and with prominent veins.

FLOWERS numerous, single on short lateral twigs, showy and large, ½ to 1½ inches (12 to 35 mm) across, bright yellow, in April and May.

SEED CAPSULE egg-shaped, 1 to 1½ inches (2.5 to 3.5 cm) long, pointed at end, 4- or 5-celled, densely hairy.

BARK deeply fissured and scaly, brownish gray, the inner bark mucilaginous.

WOOD hard, heavy, dark brown tinged with red.

HABITAT Rare and local in a few places in mountain canyons in chaparral, mainly associated with Arizona cypress, 3,200 to 6,500 feet elevation; not in New Mexico; more widespread in California.

NOTE When in flower the scattered plants are beautiful masses of yellow, standing out on the mountainside from a distance. The odd, scurfy-hairy, evergreen leaves, as well as the large bright yellow flowers, make this species an attractive ornamental.

Formerly in the Sterculia family (Sterculiaceae, now merged into Malvaceae), the same family as the cacao

CALIFORNIA FREMONTIA

tree (from which chocolate is obtained), and the Chinese parasoltree, which is planted and naturalized in the southeastern United States.

ETYMOLOGY Named in honor of *Col. John C. Fremont* (1813-90), politician, soldier, and noted explorer of the western United States, who first collected it during an 1846 expedition to California.

CALIFORNIA FREMONTIA

CALIFORNIA FREMONTIA - FRUIT

MORACEAE *Mulberry Family*

Texas Mulberry *Morus microphylla* Buckl.

ALSO CALLED **Mexican mulberry, mountain mulberry**

DESCRIPTION Small scrubby tree or large shrub to 20 feet tall and 6 inches (15 cm) in trunk diameter, often growing in dense clumps.

LEAVES in 2 rows, variable in shape, ovate and often 3- to 5-lobed, 1 to 2 inches (2.5 to 5 cm) long, short-pointed, base rounded or heart-shaped and with 3 main veins, edges coarsely saw-toothed, dark green and rough above, beneath paler and usually hairy.

MALE AND FEMALE FLOWERS in dark green catkins on different trees (dioecious), appearing with the leaves in April.

MULBERRY FRUIT an oblong cluster about ½ inch (12 mm) long, turning from red to purple or black, juicy and acid, edible.

BARK smoothish, light gray, becoming fissured and scaly.

WOOD hard, heavy, dark orange or brown.

HABITAT Common, usually along streams, canyons, washes, or rocky slopes below cliffs, in foothills and mountains, woodland and upper desert zones, 2,000 to 6,000 feet elevation, widely distributed.

NOTE The tree is widely cultivated as food for silk-worms, most species being useful for this purpose. The small mulberry fruits are eaten by Native Americans and by wildlife. This species is well-established at Grand Canyon, where the Havasupai cultivate the trees for the fruits.

ETYMOLOGY *Morus* is the Latin name.

TEXAS MULBERRY

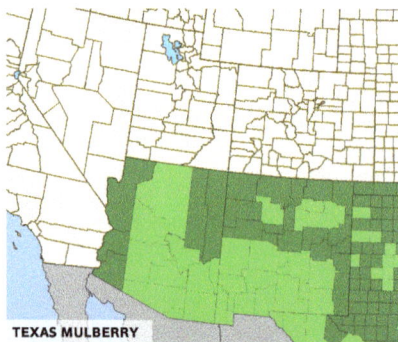

TEXAS MULBERRY

OLEACEAE *Olive Family*

Trees, shrubs, or herbs, of diverse habit; leaves simple or pinnate, alternate or opposite; flowers regular, perfect or unisexual, with or without a corolla; stamens 2 or 4; ovary 2-celled; fruit various. The best-known members of this family are olive, ash, and lilac.

1　Leaves simple; fruit not winged . . . . FORESTIERA (*Forestiera*)
1　Leaves pinnately compound or, if unifoliolate, then the blade broad, ovate, oval, or suborbicular; fruit with a conspicuous, mainly terminal, flat wing . . . . . . ASH (*Fraxinus*)

Desert-Olive Forestiera　*Forestiera shrevei* Standl.

ALSO CALLED wild-olive, desert-olive

SYNONYM *Forestiera phillyreoides* (Benth.) Torr.

DESCRIPTION Much-branched shrub 3 to 12 feet tall or sometimes a small tree as much as 20 to 25 feet tall and 8 inches (20 cm cm) in trunk diameter, evergreen or nearly so.

LEAVES paired, more or less evergreen, lance-shaped or reverse lance-shaped, ⅝ to 1 inch long (15 to 25 mm), short-pointed, edges not toothed but slightly rolled under, green and hairy on both sides.

FLOWERS male and female on different plants (dioecious), in small lateral clusters, very small, from December to March.

FRUIT egg-shaped, ¼ to ⅜ inch (6 to 9 mm) long, one-sided, resembling a small olive, with thin flesh, 1-seeded.

WOOD very hard.

HABITAT Local on dry rocky slopes and canyons in desert, often forming thickets, 2,500 to 4,500 feet elevation.

ETYMOLOGY Named for *Charles Le Forestier,* 18th-century French physician and naturalist.

DESERT-OLIVE FORESTIERA - FRUIT

DESERT-OLIVE FORESTIERA

Fraxinus ASH

Trees or large shrubs; leaves opposite, commonly pinnate, petioled; flowers in racemes or panicles, mostly unisexual, apetalous or with a 4-parted corolla; stamens commonly 2, with large anthers; fruit dry, with a large flat terminal wing, indehiscent; seeds 1 or 2. Some of the North American ashes are important timber trees, but the species occurring in the Southwest do not grow large enough to make the wood valuable. The herbage is of limited value as browse. *Fraxinus* is the classical Latin name.

1   Flowers with corolla, perfect, showy, in terminal panicles: leaves 5 to 7 inches (12.5 to 17.5 cm) long with 3 to 7 lanceolate, stalked leaflets, or the leaves sometimes simple . . . . **FRAGRANT ASH** (*Fraxinus cuspidata*)
1   Flowers without corolla, dioecious or polygamous; panicles axillary; leaves simple or pinnately compound 2

2   Leaflets ½ to ¾ inch (12 to 20 mm) long, spatulate; leaves 1½ to 3 inches (3.5 to 7.5 cm) long. . . . . **LITTLELEAF ASH** (*Fraxinus greggii*)
2   Leaflets 1 to 6 inches (2.5 to 15 cm) long; leaves (unless simple) over 3 inches (7.5 cm) long. 3

3   Twigs 4-angled; fruit compressed, oblong wing extending to base. 4
3   Twigs terete 5

4   Leaflets 1 (rarely 2 or 3); fruit ½ inch (12 mm) long . . . . **SINGLELEAF ASH** . . . . (*Fraxinus anomala*)
4   Leaflets 5 (rarely 3); fruit 1 to 1½ inches (2.5 to 3.5 cm) long. . . . . **LOWELL ASH** . . . . (*Fraxinus anomala* var. *lowellii*)

5   Leaves 9-18 inches (22 to 45 cm) long, with 7-9 leaflets . . . . **GREEN ASH** . . . . (*Fraxinus pennsylvanica*)
5   Leaves 3-7 inches (7.5 to 17.5 cm) long, with 3-5 leaflets. . . . . **VELVET ASH** (*Fraxinus velutina*)

Singleleaf Ash *Fraxinus anomala* Torr. ex S. Watson

DESCRIPTION Shrub or sometimes a small tree to 20 feet tall and 6 inches (15 cm) in trunk diameter, with rounded crown.

TWIGS 4-angled, brown, without hairs.

LEAVES paired, simple or occasionally of 2 or 3 leaflets. Leaves broadly ovate or nearly round, 1½ to 2 inches (3.5 to 5 cm) long (or leaflets 1 to 1½ inches (2.5 to 3.5 cm) long), rounded or short-pointed at tip, inconspicuously wavy-toothed or without teeth, leathery, dark green above, paler beneath, becoming hairless.

FLOWERS appearing with the leaves, in small hairy clusters, small, in April.

FRUITS long-winged "keys" ½ inch (12 mm) long, with broad rounded wing ⅜ inch (9 mm) wide extending to base; few in clusters.

BARK fissured into narrow ridges, dark brown, reddish tinged.

WOOD hard, heavy, light brown.

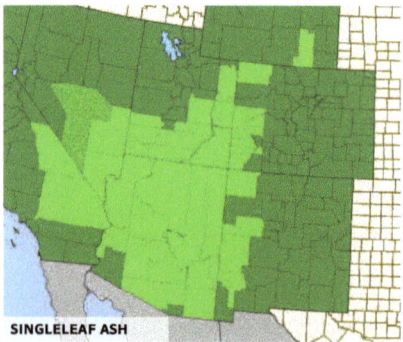

SINGLELEAF ASH

HABITAT Canyons and hillsides, often on very dry rocky slopes, in upper desert, pinyon-juniper woodland, and lower ponderosa pine forest zones, 2,000 to 6,000 feet elevation. Common at the Grand Canyon.

NOTE Singleleaf ash generally has simple leaves, making this species almost unique among the other ashes (which have pinnately compound leaves).

SINGLELEAF ASH

SINGLELEAF ASH

Lowell Ash   *Fraxinus anomala var. lowellii* (Sarg.) Little

SYNONYM *Fraxinus lowellii* Sarg.

DESCRIPTION Shrub or small tree to 25 feet in height.

TWIGS 4-angled, often winged, orange-brown the first year, becoming gray-brown.

LEAVES paired, pinnately compound, 3½ to 6 inches (8 to 15 cm) long. Leaflets 3, 5, or 7, ovate, 2¼ to 3 inches (5.5 to 7.5 cm) long, long or short-pointed, edges saw-toothed, slightly leathery, yellow-green, without hairs or slightly hairy.

FLOWERS male and female separate, in clusters ½ to 1½ inches (12 to 35 mm) long, small, in March.

FRUITS long-winged "keys" 1 to 1½ inches (2.5 to 3.5 cm) long, with broad rounded wing ⅜ inch (9 mm) or less in width and extending to base; several in clusters.

BARK deeply furrowed, brown.

HABITAT Along streams and canyons, 3,200 to 6,500 feet elevation, oak woodland and upper desert zones.

ETYMOLOGY Lowell ash commemorates *Percival Lowell* (1855-1916), American astronomer who established Lowell Observatory at Flagstaff and who found this ash in Oak Creek Canyon while collecting tree specimens in northern Arizona.

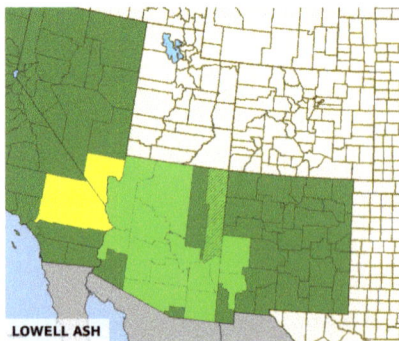

LOWELL ASH

LOWELL ASH

LOWELL ASH

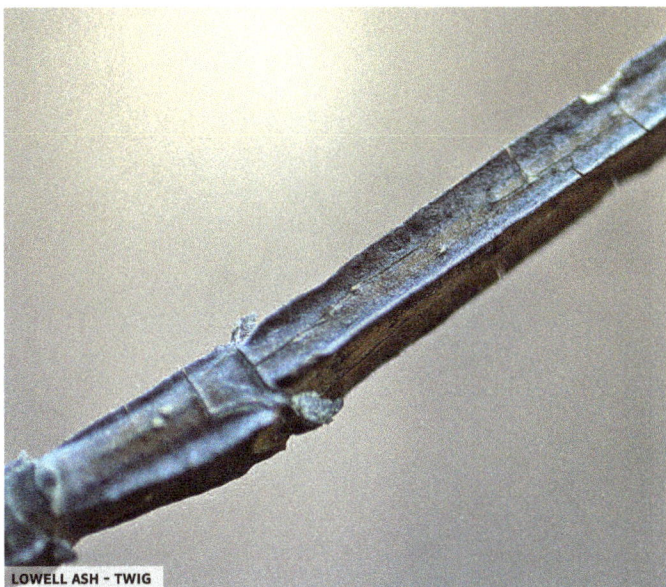

LOWELL ASH – TWIG

Fragrant Ash *Fraxinus cuspidata* Torr.

ALSO CALLED flowering ash

DESCRIPTION Shrub or small tree to 20 feet tall and 8 inches (20 cm) in trunk diameter.

LEAVES paired, pinnately compound, 3 to 7 inches (7.5 to 17.5 cm) long, or occasionally simple. Leaflets 3 to 7, long-stalked, lance-shaped or ovate, 1½ to 2½ inches (3.5 to 6 cm) long, long-pointed, slightly saw-toothed or usually without teeth, thin, shiny dark green above, beneath paler and slightly hairy when young.

FLOWERS appearing with the leaves, in hairless clusters 3 to 4 inches (7.5 to 10 cm) long, about ½ inch (12 mm) long, white, fragrant, differing from other native ashes in the larger size and presence of a 4-parted whitish corolla, in May and June.

FRUITS long-winged oblong "keys" ¾ to 1 inch (2 to 2.5 cm) long with rounded wing ¼ inch wide in upper half and extending nearly to base of flattened body; several in clusters.

BARK smoothish, gray, with age becoming much fissured into ridges and scaly.

HABITAT Scattered and local on rocky slopes of canyons and mountains in oak woodland zone, 4,500 to 7,000 feet elevation. Found along the trails into the Grand Canyon from the south rim.

NOTE Fragrant ash is planted as an ornamental because of its showy, pleasantly scented flowers and attractive foliage. The fragrant flowers having linear, showy, white tepals are unique in our ashes.

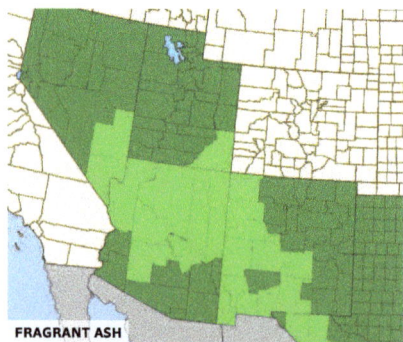

FRAGRANT ASH

FRAGRANT ASH

FRAGRANT ASH

Littleleaf Ash   *Fraxinus greggii* A. Gray

ALSO CALLED **Gregg's ash**

DESCRIPTION Shrub or sometimes a small tree to 20 feet tall and 8 inches in trunk diameter, nearly evergreen.

LEAVES paired, pinnately compound, 1 to 1½ inches (2.5 to 3.5 cm) long, axis narrowly winged, remaining attached through the winter until after flowering time. Leaflets usually 5 or 7 (or 3), reverse lance-shaped or obovate, ½ to ¾ (to 1½) inches (12 to 20 mm, to 3.5 cm) long, rounded at tip, inconspicuously toothed or without teeth, thick and leathery, dark green above, beneath paler and covered with small black dots.

FLOWERS partly male and female and partly bisexual, appearing before the leaves, in small clusters, small.

FRUITS long-winged "keys" ½ to ⅝ inch (12 to 15 mm) long with broad wing about ¼ inch (6 mm) wide extending to base and much longer than body; few in cluster.

BARK smooth, thin, gray.

WOOD hard, heavy, brown.

HABITAT On rocky slopes and in canyons, desert and oak woodland, 3,600 to 5,000 feet elevation, southwestern Texas, south into northern and central Mexico. There are also small number of collections of littleleaf ash reported from southern Arizona and New Mexico.

ETYMOLOGY *Greggii* refers to the name of its discoverer, *Josiah Gregg* (1806-50), an early American trader in the West and author of a popular book entitled *The Commerce of the Prairies.* He collected plant specimens in northern Mexico.

LITTLELEAF ASH

Chihuahuan ash *Fraxinus papillosa* Lingelsh.

Chihuahuan Ash extends northward across the Mexican border into several mountainous areas in southeastern Arizona and southwestern New Mexico. It is recognized by the whitish lower surfaces of the leaflets; under a microscope, the lower surfaces of the leaves appear as a solid mass of whitish beads, which are the tiny projections (or papillae) of each epidermal cell. It is distinguished from the region's most common ash, **Velvet Ash** (*Fraxinus velutina*), by its leathery, more or less sessile leaflets, whch are somewhat whitened on their undersurface.

ADDITIONAL SPECIES Green Ash (*Fraxinus pennsylvanica* Marsh.), a medium-sized tree, native to the eastern and central USA but considered adventive in Arizona, New Mexico, and the western states. Leaves paired, pinnately compound; leaflets 5 to 9 (usually 7), each 3 to 4 inches (7.5 to 10 cm) long, 1 to 2 inches wide (2.5 to 5 cm) broad, the short stalk narrowly winged. Fruit a samara with a flat wing arising usually in the upper ¼ of the seed body. Found as an escape from plantings.

CHIHUAHUAN ASH

GREEN ASH

Velvet Ash *Fraxinus velutina* Torr.

ALSO CALLED desert ash, smooth ash, Arizona ash, Toumey ash, fresno

SYNONYM *Fraxinus standleyi* Rehd.

DESCRIPTION Small to medium-sized tree to 40 feet tall and 1 foot or more in trunk diameter, with spreading branches and a rounded crown.

TWIGS brown, hairy or without hairs.

LEAVES paired, pinnately compound, 3 to 6 inches (7.5 to 15 cm) long. Leaflets 5 to 9, varying greatly in appearance, lance-shaped or elliptical, 1 to 3 inches (2.5 to 7.5 cm) long, short or long-pointed, edges slightly toothed or without teeth, varying from thin to thick and leathery, and varying from densely short-hairy beneath to hairless.

FLOWERS yellow (male) and green (female) on different trees (dioecious), appearing before the leaves, many in clusters, small, in March and April.

FRUITS long-winged "keys" ¾ to 1 inch (2 to 2.5 cm) long, wing not extending to base; many in dense clusters.

BARK deeply furrowed into broad ridges, gray.

WOOD soft, heavy, light brown.

HABITAT Common and characteristic tree of streambanks, moist washes, and moist canyons, in the desert, desert grassland, oak woodland, and ponderosa pine forest zones, 2,500 to 7,000 feet elevation.

NOTE Highly variable in characters of leaflets, such as shape, edges, thickness, hairiness, and length of leaflet stalks. The species is popular as a shade tree and is widely planted in southern Arizona and California.

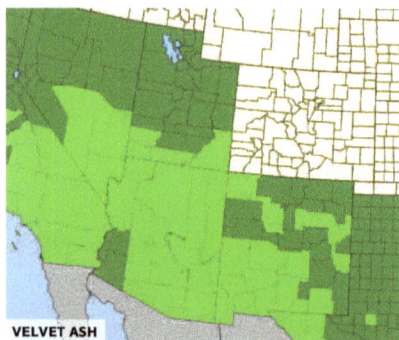

VELVET ASH

VELVET ASH

VELVET ASH

VELVET ASH

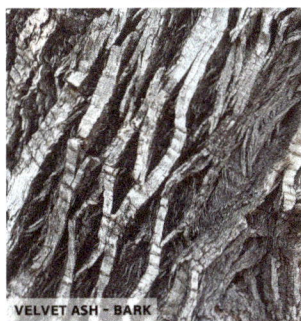

VELVET ASH - BARK

PLATANACEAE *Sycamore Family*

Arizona Sycamore *Platanus wrightii* S. Wats.

ALSO CALLED Arizona planetree

SYNONYM *Platanus racemosa* var. *wrightii* (S. Wats.) L. Benson

DESCRIPTION Large tree 40 to 80 feet in height, with large trunk to 4 feet or more in diameter, and with large spreading branches, and a broad open crown.

LEAVES slightly star-shaped, large, 6 to 10 inches (15 to 25 cm) long and wide, deeply divided into usually 5 or 3 (or sometimes 7) narrow, long-pointed lobes, the edges mostly without teeth, thin and firm, light green and hairless above, beneath pale and hairy. Leafstalks 1½ to 3 inches (3.5 to 7.5 cm) long, stout, the enlarged base enclosing the bud.

FLOWERS male and female separate on the same tree; female flowers in ball-like clusters (these in groups of 2 to 4), in March and April.

FRUIT of 2 to 4 stalked balls, ¾ to 1 inch (2 to 2.5 cm) in diameter, along a stalk 6 to 8 inches (15 to 20 cm) long.

BARK of branches smooth, thin, whitish; bark of largest branches and trunk smoothish and peeling off in brownish flakes, on large trunks becoming thick, dark gray, and deeply and irregularly divided into thick plates.

WOOD light brown.

HABITAT Common along streams and rocky canyons in foothills and mountains, upper desert, desert grassland, and oak woodland zones; 2,000 to 6,000 feet elevation.

NOTE Arizona sycamore, one of the largest and handsomest deciduous tree species in Arizona, is valuable in preventing erosion along stream banks.

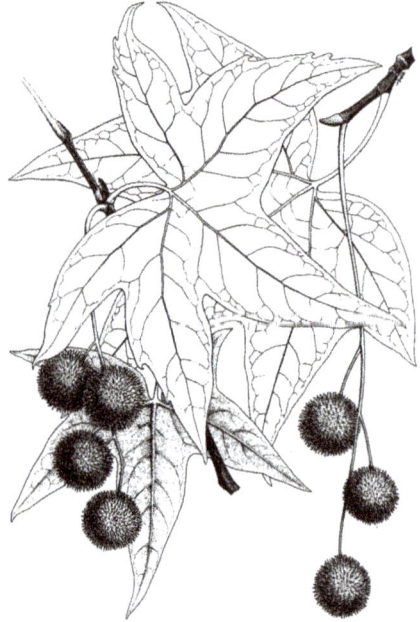

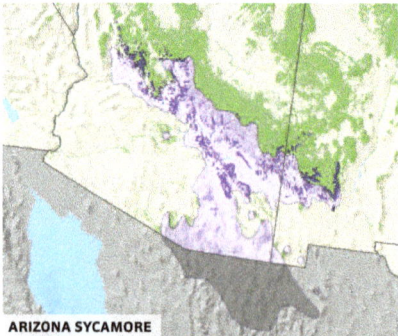

ARIZONA SYCAMORE

ARIZONA SYCAMORE

ETYMOLOGY *Platanus,* the Greek name. The species name honors *Charles Wright,* 1811-85, botanical collector mainly in Texas, Cuba and Connecticut.

ARIZONA SYCAMORE – BARK

ARIZONA SYCAMORE – FRUIT

RHAMNACEAE *Buckthorn Family*

Shrubs or small trees; leaves simple; flowers small, perfect or unisexual, regular or nearly so, 4- or 5-merous, with or without petals, with a disk in the calyx throat on which the stamens often are borne; ovary 2- or 3-celled, superior or partly inferior. Most of our species are browsed by domestic animals and deer.

1 Fruit dry or somewhat berry-like, 2- or 3-seeded . . . . **BUCKTHORN** (*Rhamnus*)
1 Fruit fleshy, black, with a 1- to 3-celled stone. 2

2 Leaves elliptic to ovate, commonly at least ¼ inch (6 mm) wide; petals present; fruit globose (or nearly so), not beaked, ¼ to ⅜ inch (6 to 8 mm) in diameter, dark blue with a waxy bloom; young stems glaucous . . . . **LOTEBUSH** (*Zizyphus*)
2 Leaves spatulate; petals none; fruit ovoid, ellipsoid, or somewhat obovoid, often beaked with the persistent style, less than ¼ inch (6 mm) in diameter, at maturity black without a waxy bloom; young stems not glaucous . . . . **CONDALIA** (*Condalia*)

Bitter Condalia *Condalia globosa* I.M. Johnst.

DESCRIPTION Spiny large shrub or small tree as much as 15 to 20 feet tall and 1 foot or more in trunk diameter, with broad crown of tangled spreading branches.

TWIGS stiff, light gray or brown, many ending in sharp spines.

LEAVES relatively few, not crowded, single or clustered, very small, oblong but narrowed toward base, ¼ to ½ inch (6 to 12 mm) long, edges not toothed, usually densely and finely hairy.

FLOWERS 1 or 2 in leaf axils, small.

FRUIT berrylike, almost ¼ inch (6 mm) in diameter, on stalks about the same length, black, juicy, very bitter, 1-seeded.

BARK much fissured and becoming shreddy, brownish gray.

HABITAT Dry sandy plains and along washes in desert, 1,000 to 2,500 feet elevation; not in New Mexico.

NOTE Originally described in 1924 from Lower California, Mexico, as a shrub 4 to 8 feet tall, bitter condalia becomes a small tree in southwestern Arizona. Very similar to *Zizyphus*, but the

BITTER CONDALIA

leaves spatulate and finely pubescent, and the petals absent.
ETYMOLOGY Named for *Antonio Condal,* an 18th Spanish physician.

BITTER CONDALIA

BITTER CONDALIA

Rhamnus BUCKTHORN

Shrubs or small trees, not spiny; leaves opposite; flowers perfect or unisexual, greenish, axillary, in small fascicles or solitary, with or without petals; calyx free from the ovary; fruit a drupe with 2 to 4 stones.

All of our species have ornamental value. The laxative drug cascara is obtained from the bark of *Rhamnus purshiana,* a species of the Pacific Coast. The plants, especially the more or less evergreen forms, are of some value as browse in winter, and the fruits are eaten by birds. *Rhamnus* is the ancient Greek name.

1   Bud scales present; flowers in fascicles without a common peduncle, commonly 4-merous; style exserted; leaves evergreen, often bronzed beneath; fruits bright red at maturity . . . . **HOLLYLEAF BUCKTHORN** (*Rhamnus ilicifolia*)
1   Bud scales absent; flowers in pedunculate cymes, commonly 5-merous; style not exserted; leaves evergreen or deciduous; fruits black or nearly so at maturity 2
2   Leaves evergreen, thickish, whitish-tomentose beneath, oblong-lanceolate to broadly elliptic, usually less than 1¼ (3 cm) wide; fruits usually 2-seeded . . . . **CALIFORNIA BUCKTHORN** (*Rhamnus californica*)
2   Leaves deciduous, thin, green on both faces, sparsely to copiously pubescent beneath but not tomentose, broadly elliptic to ovate-oblong, usually more than 1¼ (3 cm) wide; fruits usually 3-seeded. . . . . **BIRCHLEAF BUCKTHORN** (*Rhamnus betulifolia*)

Birchleaf Buckthorn   *Rhamnus betulifolia* Greene

SYNONYM *Frangula betulifolia* (Greene) Grubov
DESCRIPTION Rounded shrub 8 feet or less in height, branching from the base, or rarely a small tree to 18 feet tall and 4 inches (10 cm) or more in trunk diameter.
LEAVES deciduous, oblong to elliptical or, in var. *obovata,* obovate, 2 to 4 inches (5 to 10 cm) long, blunt or short-pointed at tip, with edges finely and sharply toothed and not rolled under, thin, bright green, slightly hairy beneath.
FLOWERS several in small cluster at leaf axil, small, less than ⅛ inch long, greenish, in May and June.
BERRIES ⅜ inch (9 mm) in diameter, juicy, usually 3-seeded.

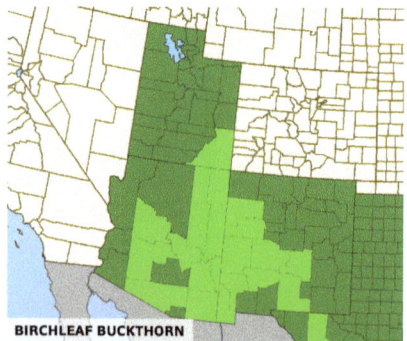

BIRCHLEAF BUCKTHORN

HABITAT Canyons and along streams in mountains, usually in shade, oak woodland and ponderosa pine forest zones, 5,500 to 7,500 feet elevation.
NOTE Birchleaf buckthorn is browsed by deer and bighorns. The Apache ate the fruits with meat.

BIRCHLEAF BUCKTHORN

BIRCHLEAF BUCKTHORN

California Buckthorn *Rhamnus californica* subsp. *ursina* (Greene) C.B. Wolf

ALSO CALLED coffeeberry, pigeonberry

SYNONYM *Frangula californica* subsp. *ursina* (Greene) Kartesz & Gandhi, *Rhamnus ursina* Greene

DESCRIPTION Evergreen shrub 6 feet or more in height, commonly branched from base but sometimes a small tree to 20 feet tall and 6 inches (15 cm) in trunk diameter, with rounded spreading crown.

LEAVES elliptical or oval, 1¼ to 3 inches (3 to 7.5 cm) long, short-pointed at tip and base, edges finely and inconspicuously toothed and slightly rolled under, leathery, above dull green and hairless or nearly so, beneath paler and densely hairy.

FLOWERS several in small cluster at leaf axil, small, less than ⅛ inch (3 mm) long, greenish, from May to July.

BERRIES changing from green to red and to black at maturity, ⅜ inch (9 mm) in diameter, juicy, with 2 or 3 large seeds.

BARK smooth, gray.

HABITAT Common in canyons and along streams in mountains in chaparral, oak woodland, pinyon-juniper woodland, and ponderosa pine forest zones, 3,500 to 7,000 feet elevation.

NOTE California buckthorn generally is a shrub, rather than a tree, and has several varieties besides the southwestern one described here. The typical form found along the Pacific Coast in California and Oregon has the leaves hairless or nearly so beneath.

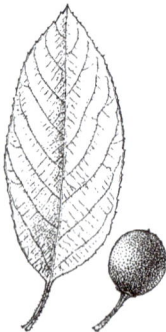

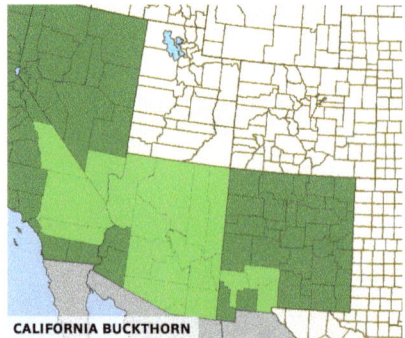

CALIFORNIA BUCKTHORN

CALIFORNIA BUCKTHORN

CALIFORNIA BUCKTHORN

Hollyleaf Buckthorn *Rhamnus ilicifolia* Kellogg

ALSO CALLED **hollyleaf redberry buckthorn**

SYNONYM *Rhamnus crocea* var. *ilicifolia* (Kellogg) Greene

DESCRIPTION Evergreen shrub or small tree to 15 feet in height, with spreading branches.

LEAVES hollylike, oval to nearly round, ¾ to 1½ inches (2 to 3.5 cm) long, rounded at tip and base, edges spiny-toothed, leathery, shiny yellow-green above, beneath yellow-green or paler and hairless or nearly so.

FLOWERS usually male and female on different plants, few in small clusters in leaf axils, small, about ⅛ inch (3 mm) long, yellow-green, from March to May.

BERRIES bright red, ¼ inch (6 mm) in diameter, juicy, usually 2-seeded.

BARK slightly rough and fissured, dark gray.

WOOD light brown.

HABITAT Common in mountains in chaparral and lower ponderosa pine forest, 3,000 to 7,000 feet elevation; not in New Mexico.

NOTE Though generally a shrub in Arizona, hollyleaf buckthorn is sometimes a small tree with a distinct trunk. It is an evergreen component of the California chaparral vegetation which reappears in the mountains of Arizona. As the spiny evergreen leaves and red berries resemble holly, this attractive shrub is sometimes used for Christmas decorations.

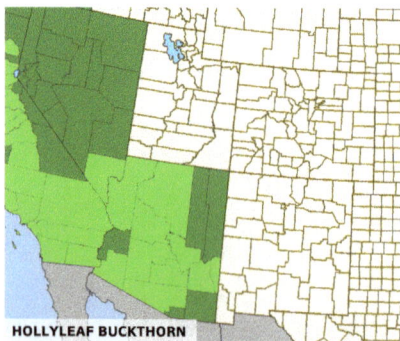

HOLLYLEAF BUCKTHORN

HOLLYLEAF BUCKTHORN

HOLLYLEAF BUCKTHORN

HOLLYLEAF BUCKTHORN

Lotebush *Ziziphus obtusifolia* (Hook. ex Torr. & A. Gray) A. Gray

ALSO CALLED **lotewood condalia, white crucillo, gray-thorn**

SYNONYM *Condalia lycioides* (A. Gray) Weberb., *Condalia obtusifolia* (Hook.) Weberb., *Sarcomphalus obtusifolius* (Hook. ex Torr. & Gray) Hauenschild, *Zizyphus lycioides* A. Gray

DESCRIPTION Spiny, much-branched shrub 3 to 8 feet tall, rarely 10 feet tall with a trunk 4 inches (10 cm) in diameter or larger.

TWIGS numerous and widely spreading at right angles, 1 to 4 inches (2.5 to 10 cm) long, light gray, hairless or in a variety densely short-hairy, ending in sharp spines.

LEAVES small elliptical to ovate, ⅜ to ¾ inch (9 to 20 mm) long and ¼ inch (6 mm) broad, hairless (or in a variety densely short-hairy), pale green.

FLOWERS several in a slightly stalked cluster in leaf axil, small, less than ⅛ inch long, whitish green.

FRUIT berrylike, more than ¼ inch (6 mm) in diameter, blue black with a bloom, juicy and sweet, 1-seeded.

HABITAT Common, often forming thickets, on plains and mesas in desert and desert grassland, 1,000 to 5,000 feet elevation.

NOTE This shrubby species rarely becomes a small tree. A variety in Arizona (except in southeastern Cochise County), California, and adjacent Mexico, has densely short-hairy twigs and leaves.

ETYMOLOGY The generic name is derived from *zizfum* or *zizafun,* the Persian word for the plant *Ziziphus lotus.*

LOTEBUSH

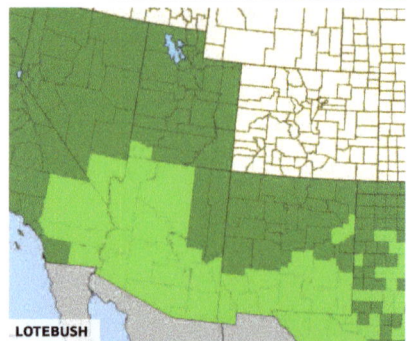

LOTEBUSH

ROSACEAE *Rose Family*

Plants herbaceous or woody; leaves alternate, simple or compound, usually with stipules; flowers mostly perfect, regular or nearly so; sepals partly united; petals commonly 5, occasionally none; stamens commonly numerous, rarely fewer than 5, nearly always borne on the throat of the calyx or on a disk surrounding the ovary or ovaries; pistils 1 to many; ovary free from or adnate to the calyx; fruit various.

This large and very diverse family includes many of the most important cultivated fruits, such as the apple, pear, peach, plum, cherry, apricot, almond, strawberry, raspberry, and blackberry. The flowers and foliage are usually attractive, often beautiful, and many plants of this family, first and foremost the roses, are highly prized as cultivated ornamentals. Many of our species are important browse plants, both for domestic animals and deer, and the fruits supply much food for birds and other wild animals.

1   Carpel solitary; fruit a dry or fleshy, usually 1-seeded drupe (plumlike); calyx more or less persistent at base of the fruit; plants small trees or large shrubs; leaves simple; flowers white or greenish, in racemes or corymbs, or solitary in the leaf axils . . . .
. . . . **CHOKECHERRY, PLUM** (*Prunus*)
1   Carpels more than one or, if solitary, the fruit an achene 2

2   Ovary inferior, enclosed in and joined to the calyx tube (hypanthium) , the latter becoming more or less fleshy; fruit a pome (applelike); calyx lobes more or less persistent at apex of the fruit; plants shrubby or treelike; petals white. 3
2   Ovary superior; calyx tube not fleshy and enclosing the pistils or, if so, then not joined to them. 4

3   Plants unarmed; ovary with complete and false partitions, the cells twice as many as the number of styles; flowers relatively large, in racemes or corymbose fascicles. . . . .
. . . . **SERVICEBERRY** (*Amelanchier*)
3   Plants armed with strong spines; ovary without false partitions, the cells of the same number as the styles; flowers relatively small, in corymblike cymes . . . . . . **HAWTHORN**
. . . . (*Crataegus*)

4   Carpels becoming dehiscent capsules or follicles, containing usually more than one seed. . . . . **VAUQUELINIA** (*Vauquelinia*)
4   Carpels becoming indehiscent 1-seeded achenes 5

5   Petals normally none; leaf blades entire or merely dentate . . **MOUNTAIN-MAHOGANY**
. . . . (*Cercocarpus*)
5   Petals present; leaf blades wedge-shaped, usually deeply cleft or pinnatifid . . . .
. . . . **CLIFFROSE** (*Purshia*)

Utah Serviceberry *Amelanchier utahensis* Koehne

SYNONYM *Amelanchier australis* Standl., *A. goldmanii* Woot. & Standl., *A. mormonica* Schneid., *A. rubescens* Greene

DESCRIPTION Shrub or small tree 5 to 16 feet tall, much-branched and often forming clumps.

LEAVES very variable, nearly round to elliptic, small, ½ to 1¼ inches (12 to 30 mm) long, rounded at tip and base, edges coarsely and often sharply toothed above middle, slightly leathery at maturity, grayish green, finely hairy on both sides.

FLOWERS 3 to 6 in a cluster, about ½ inch (12 mm) long, with 5 white petals, from April to June.

FRUIT ¼ to ⅜ inch (6 to 9 mm) in diameter, resembling a small apple, bluish black, juicy and sweet, or often pale brown, dry and mealy.

BARK smooth, ashy gray.

HABITAT Scattered in canyons, rocky slopes, foothills, and mountains in pinyon-juniper woodland and ponderosa pine forest and sometimes also in upper desert, 2,000 to 8,500 feet elevation.

NOTE The fruits are eaten by wildlife and were consumed by the Native Americans fresh or dried, though in arid localities the fruits may remain dry and insipid and not become juicy. The foliage is good to excellent browse for sheep and cattle.

ETYMOLOGY From *amelancier,* the French Provençal name of one of the species. It is also called "sarvis" or "servis berry" from its resemblance to the service, a forgotten English fruit rather like a pear.

UTAH SERVICEBERRY

UTAH SERVICEBERRY

UTAH SERVICEBERRY - FRUIT

UTAH SERVICEBERRY

Cercocarpus MOUNTAIN-MAHOGANY

Shrubs or small trees; leaves simple, fascicled, with thickish, entire or dentate blades, these linear to obovate, often prominently veined beneath; flowers solitary or in small fascicles, inconspicuous, with small yellowish sepals and no petals; stamens numerous; hypanthium sheathlike in fruit, enclosing the slender villous achene, the long, persistent, plumose style exserted.

The plants are sometimes known locally as "deerbrowse," and certain species are important elements of the chaparral in central and southern Arizona, useful in protecting the soil against erosion and affording excellent browse for cattle, sheep, and goats, as well as for deer. Cases have been reported of hydrocyanic-acid poisoning of animals eating the leaves of *C. montanus.* The wood is hard and that of some species was used by Native Americans for making digging sticks and is occasionally used for making tool handles. The sharp-pointed basal end of the achene and the corkscrew-like tail enable it to penetrate the ground, as in the needle-grasses. *Cercocarpus* is from the Greek *kerkos,* tail and *karpos,* fruit.

1   Leaves persistent, leathery, resinous, linear or elliptic, acute at both ends, entire, the lateral veins not very prominent on leaf underside. . . . . **CURLLEAF MOUNTAIN-MAHOGANY** (*Cercocarpus ledifolius*)
1   Leaves deciduous, often thickish but scarcely leathery, not noticeably resinous, elliptic to broadly obovate,, the lateral veins very prominent on leaf underside **2**

2   Blades entire or toothed only at or very near the tip, commonly oblanceolate or spatulate. . . . . **HAIRY MOUNTAIN-MAHOGANY** (*Cercocarpus breviflorus*)
2   Blades toothed well below the tip, mostly obovate, commonly 2 to 3 times as long as wide . . . . **BIRCHLEAF MOUNTAIN-MAHOGANY** (*Cercocarpus betuloides*)

Birchleaf Mountain-Mahogany   *Cercocarpus betuloides* Nutt.

SYNONYM *Cercocarpus montanus* var. *glaber* (S. Wats.) F.L. Martin.
DESCRIPTION Deciduous, large spreading shrub, or small tree to 20 feet in height with a single trunk to 6 inches (15 cm) in diameter and spreading crown.
LEAVES obovate to oval, 1 to 1¼ inches (2.5 to 3 cm) long and ⅜ to ½ inch (9 to 12 mm) wide, rounded at tip and wedge-shaped at base, finely toothed above middle, slightly thick, dark green above, pale green or grayish and slightly hairy beneath.
FLOWERS 1 to 3 in leaf axils, about ½ inch (12 mm) long, yellowish, March to July.
FRUIT hairy, with a twisted hairy tail (style) 2 to 3 inches (5 to 7.5 cm) long.
BARK of thin scales, smoothish.
HABITAT Chaparral zone of mountains in Arizona, 3,500 to 6,500 feet elevation.

BIRCHLEAF MOUNTAIN-MAHOGANY

Hairy Mountain-Mahogany *Cercocarpus breviflorus* A. Gray

ALSO CALLED **Eastern mountain-mahogany, Wright mountain-mahogany**

SYNONYM *Cercocarpus montanus* var. *paucidentatus* (S. Wats.) F. L. Martin, *Cercocarpus paucidentatus* (S. Wats.) Britton

DESCRIPTION Deciduous shrub or small tree to 15 feet or more in height and 5 inches (12.5 cm) in trunk diameter, with open crown of widely spreading branches and long slender twigs.

LEAVES obovate or reverse lance-shaped, ⅜ to 1 inch (9 to 25 mm) long and ¼ to ½ inch (6 to 12 mm) wide, rounded or short-pointed at tip and tapering from above middle to base, edges turned under and usually with a few rounded teeth near tip, thick, dark green and hairless or slightly hairy above, pale and finely hairy beneath.

FLOWERS 1 to 3 in leaf axils, about ½ inch (12 mm) long, yellowish, March to November.

FRUIT ¼ inch (6 mm) long, red-brown, hairy, with a twisted white-hairy tail (style) 1 to 1½ inches (2.5 to 3.5 cm) long.

BARK thin, smoothish, becoming fissured and scaly, gray or red-brown.

WOOD hard, light brown with light yellow sapwood.

HABITAT Common on dry slopes and mountainsides, chaparral and oak woodland, 5,000 to 8,000 feet elevation.

NOTE This species and birchleaf mountain-mahogany are important browse plants for livestock.

HAIRY MOUNTAIN-MAHOGANY

HAIRY MOUNTAIN-MAHOGANY

Curlleaf Mountain-Mahogany *Cercocarpus ledifolius* Nutt.

DESCRIPTION Evergreen spreading shrub or small tree to 15 or sometimes 20 feet tall and ½ to 2 feet in trunk diameter, with compact rounded crown of widely spreading branches and many stiff twigs.

LEAVES usually clustered, narrowly lance-shaped or elliptic, ½ to 1¼ inches (12 to 30 mm) long and less than ⅜ inch (9 mm) wide, short-pointed at both ends; edges without teeth, rolled under; thick and leathery, shiny dark green above, pale and finely hairy beneath, resinous and slightly aromatic.

FLOWERS single in leaf axils, ⅝ inch (15 mm) long, yellowish, in April.

FRUIT ¼ inch (6 mm) long, hairy, with a twisted hairy tail (style) 1½ to 3 inches (3.5 to 7.5 cm) long.

BARK thick, furrowed and scaly, reddish brown.

WOOD very hard, heavy, red or dark brown with thin yellow sapwood.

HABITAT Common locally on both rims of Grand Canyon, in ponderosa pine forest zone, about 7,000 feet elevation; rare in northwestern New Mexico.

CURLLEAF MOUNTAIN-MAHOGANY

CURLLEAF MOUNTAIN-MAHOGANY

CURLLEAF MOUNTAIN-MAHOGANY

Crataegus HAWTHORN

Shrubs or small trees, armed with strong sharp thorns; leaves simple, petioled, serrate to shallowly lobed, strongly veined, sparsely pubescent beneath or glabrate; flowers in several-flowered corymbs; petals rather small, round, usually white; fruits nearly globose, thin-fleshed, nearly filled by the large bony seeds. *Crataegus* **is the Greek name for the tree; from** *kratos,* **strength; an allusion to the strength and hardness of the wood.**

ADDITIONAL SPECIES Fireberry hawthorn (*Crataegus chrysocarpa* Ashe) is reported from Colfax County, New Mexico, becoming more common north and east of our region. Plants a spiny shrub or small tree; twigs hairy when young, with many spines ¾ to 2½ inches (2 to 6 cm) long; leaves broadly ovate or nearly round, 1¼ to 2 inches (3 to 5 cm) long, short-pointed, broadly wedge-shaped at base, slightly 7- or 9-lobed, coarsely and doubly saw-toothed, firm, dark green above and paler beneath, loosely hairy on both sides; flowers in clusters, white, ¾ inch (2 cm) or less across; fruit about ⅜ inch (9 mm) in diameter, orange or red, resembling a small apple, with thin flesh and bony nutlets.

Wooton's hawthorn (*Crataegus wootoniana* Eggleston), is a shrub 10 feet or less in height, occurs in the White Mountains and Mogollon Mountains of southern and southwestern New Mexico. It has broadly ovate, saw-toothed leaves with 3 or 4 pairs of broad lobes.

1 Spines few, not more than 1 inch (2.5 cm) long; leaf blades elliptic, about twice as long as wide, not or scarcely lobed, tapering at base .... **RIVER HAWTHORN** .... (*Crataegus rivularis*)
1 Spines numerous, 1¼ to 2 inches (3 to 5 cm) long; leaf blades ovate, less than twice as long as wide, often distinctly lobed, rather abruptly contracted at base .... **CERRO HAWTHORN** (*Crataegus erythropoda*)

Cerro Hawthorn *Crataegus erythropoda* Ashe

ALSO CALLED manzana de puya larga
SYNONYM *Crataegus cerronis* A. Nels.
DESCRIPTION Spiny shrub or small tree to 15 feet in height.
TWIGS more or less zigzag, with many stout sharp purplish spines ¾ to 2 inches (2 to 5 cm) long, without hairs, purplish.
LEAVES ovate, 1 to 2½ inches (2.5 to 6 cm) long, less than twice as long as wide, short-pointed, wedge-shaped at base, often shallowly 3- to 7-lobed, coarsely saw-toothed, bright green and without hairs.
FLOWERS in clusters 1 to 2½ inches (2.5 to 6 cm) broad, white.
FRUIT about ⅜ inch (9 mm) in diameter, orange-red, resembling a small apple, with thin flesh and large bony nutlets.
HABITAT Uncommon along streams and canyons, pinyon-juniper woodland and ponderosa pine forest, 5,400 to 8,000 feet elevation.

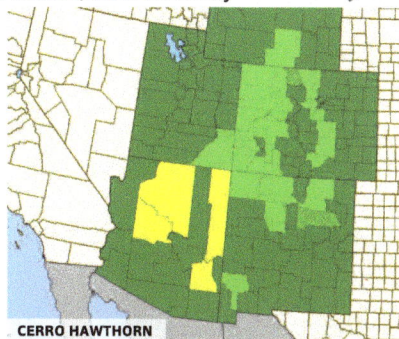

CERRO HAWTHORN

River Hawthorn *Crataegus rivularis* Nutt.

ALSO CALLED black haw

DESCRIPTION Spiny shrub or small tree to 20 feet tall, with open crown, often forming clumps.

TWIGS more or less zigzag, shiny reddish brown, often with a few straight, slender, sharp spines 1 inch (2.5 cm) or less in length.

LEAVES elliptic, about 2 inches (5 cm) long and less than 1 inch (2.5 cm) wide, mostly short-pointed, tapering at base, finely saw-toothed but not lobed, thin, without hairs at maturity, blue-green above, pale yellow-green beneath.

FLOWERS several in cluster, ½ inch (12 mm) across, white.

FRUIT ⅜ to ½ inch (9 to 12 mm) in diameter, dark red but turning black, shiny, resembling a small apple, with thin mealy flesh and 3 to 5 large bony nutlets.

HABITAT Local and scattered along streams and canyons in mountains, pinyon-juniper woodland and ponderosa pine forest zones, 5,000 to 6,000 feet elevation.

RIVER HAWTHORN

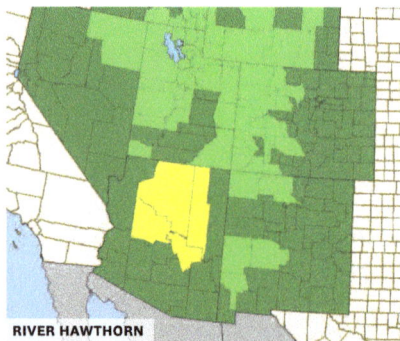

RIVER HAWTHORN

Prunus CHERRY

Small trees or large shrubs; leaves alternate or fascicled, simple; flowers usually perfect, in racemes or corymbs, or solitary in the leaf axils; calyx free from the ovary, 5-lobed, with a disk at the bottom; petals 5, these and the numerous stamens inserted on the calyx; pistil 1, the ovary 1-celled; fruit a dry or fleshy drupe, with 1 bony seed. The plants are browsed, but cases have been reported of hydrocyanic acid poisoning of cattle and sheep. The fruits are much eaten by birds and animals, and preserves are sometimes made from chokecherries. *Prunus* is the Latin name.

1   Fruit usually over ½ inch (12 mm) in diameter and usually slightly 2-lobed by ventral groove; terminal bud absent.................. AMERICAN PLUM (*Prunus americana*)
1   Fruit seldom ½ inch (12 mm) across, not lobed; leaves folded in bud; terminal bud present..2

2   Flowers few, in umbels or corymbs .... BITTER CHERRY (*Prunus emarginata*)
2   Flowers 12 or more in elongated racemes 3

3   Calyx deciduous from fruit; leaves oblong-oval to obovate, with spreading teeth ....
    .... COMMON CHOKECHERRY (*Prunus virginiana*)
3   Calyx persistent on fruit.......................... SOUTHWESTERN CHOKECHERRY
    .... (*Prunus serotina* var. *rufula*)

American Plum   *Prunus americana* Marsh.

ALSO CALLED wild plum

DESCRIPTION Shrub or small tree to 10 feet or more in height, sometimes slightly spiny, often forming dense thickets.

TWIGS stout and stiff, widely spreading, the short lateral twigs often ending in spines.

LEAVES oval, 2 to 4 inches (2.5 to 5 cm) long, long-pointed, narrowed at base, sharply saw-toothed, thick and firm, appearing slightly wrinkled, dark green above and paler beneath, hairless or nearly so.

FLOWERS 2 to 5 in a cluster, ¾ to 1 inch (2 to 2.5 cm) across, white, in April before the leaves.

FRUIT a plum about ¾ inch (2 cm) in diameter, yellowish red, juicy, acid, and edible, with a large stone.

BARK gray.

WOOD hard and heavy, dark brown tinged with red.

HABITAT Moist soil along streams, ditches, and hillsides and as an escape from cultivation, 5,000 to 7,200 feet elevation.

NOTE The fruits of this wild plum are widely used in making preserves and jellies. As the plants form dense thickets they are excellent for erosion control and have been grown in quantities for this purpose. Over 200 named varieties of this native plum have been selected for cultivation.

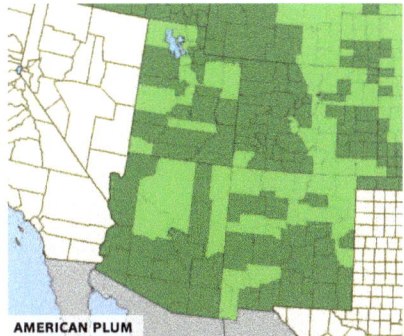

AMERICAN PLUM

Bitter Cherry *Prunus emarginata* (Dougl.) Eaton

ALSO CALLED wild cherry

DESCRIPTION Small tree or shrub to 13 feet or more in height, with slender upright branches.

TWIGS red, slender, hairy when young.

LEAVES varying in shape, oval or elliptic to reverse lance-shaped, 1 to 2 inches (2.5 to 5 cm) long, rounded or short-pointed at tip, edges minutely saw-toothed with blunt gland-tipped teeth, dark green above, pale and sometimes hairy beneath, very bitter.

FLOWERS few in short clusters 1 to 1½ inches (2.5 to 3.5 cm) long, ⅜ to ½ inch (12 mm) across, white, April to June.

FRUIT a red cherry about ⅜ inch (9 mm) in diameter, juicy, acid and bitter, with a large stone.

BARK smooth, purplish or reddish brown, very bitter.

WOOD soft and brittle, brown streaked with green.

HABITAT Mountains in ponderosa pine forest, 5,000 to 9,000 feet elevation.

BITTER CHERRY

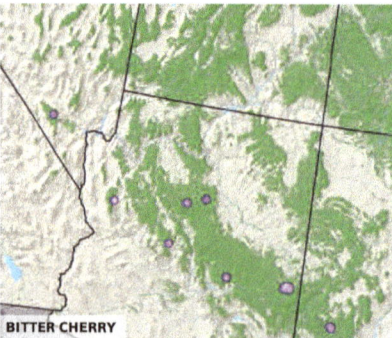

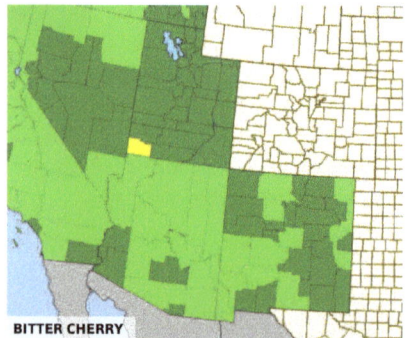

BITTER CHERRY

BITTER CHERRY

Southwestern Chokecherry *Prunus serotina var. rufula*
(Wooton & Standl.) McVaugh

ALSO CALLED Gila chokecherry, Chisos wild cherry

SYNONYM *Prunus virens* (Woot. & Standl.) Shreve, *Prunus rufula* Woot. & Standl.

DESCRIPTION Small to medium-sized tree to 40 feet in height and 2 feet in trunk diameter, with spreading crown, or a large shrub.

LEAVES elliptic, 1½ to 2 inches (3.5 to 5 cm) long, short-pointed at tip and base, edges finely saw-toothed, shiny light green above, beneath paler and without hairs or with rusty brown hairs along midrib.

FLOWERS many in clusters 3 to 6 inches (7.5 to 15 cm) long, small, ¼ inch (6 mm) across, white, in April and May.

FRUIT a chokecherry about ⅜ inch (9 mm) in diameter, with calyx remaining attached at base, purplish black and shiny, juicy but astringent, with a large stone.

BARK fissured into thin flat plates, gray, on larger trunks rough and scaly, black.

HABITAT Common, usually along streams and in canyons, mountains in oak woodland zone, 4,500 to 7,500 feet elevation.

SOUTHWESTERN CHOKECHERRY

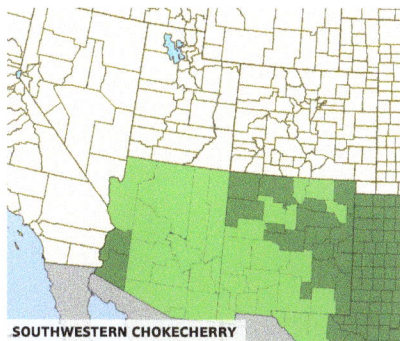

SOUTHWESTERN CHOKECHERRY

Common Chokecherry *Prunus virginiana* L.

ALSO CALLED wild cherry

SYNONYM *Padus valida* Woot. & Standl.

DESCRIPTION Shrub or small tree to 25 feet in height, with a trunk 6 inches (15 cm) or more in diameter, often forming dense thickets.

LEAVES varying in shape and hairiness, oval to obovate, 2 to 4 inches (5 to 10 cm) long, abruptly short-pointed at tip, rounded or heart-shaped at base, edges sharply saw-toothed, thick, above shiny dark green, beneath light green and without hairs or slightly hairy.

FLOWERS many in clusters 3 to 6 inches (7.5 to 15 cm) long, small, ⅜ to ½ inch (9 to 12 mm) across, white, in April and May.

FRUIT a chokecherry ¼ to ⅜ inch (6 to 9 mm) in diameter, at maturity dark red or nearly black, shiny, juicy and astringent, with a large stone.

BARK smooth and brown or gray on small trunks, becoming fissured and scaly.

WOOD hard and heavy, light brown.

HABITAT Widely distributed, especially along streams in mountains, ponderosa pine forest, 4,500 to 8,000 feet elevation. This species has a very broad range from Newfoundland and Quebec west to British Columbia and south to California, Texas, and Georgia.

NOTE Chokecherries are eaten by wildlife and are sometimes used in making preserves and jelly. However, as the common name suggests, they are astringent when raw.

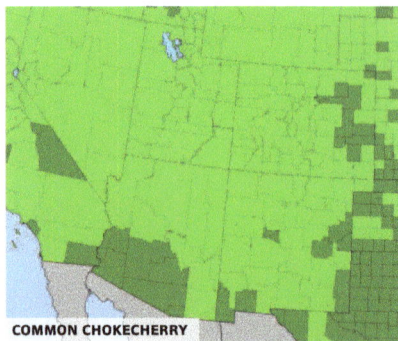

COMMON CHOKECHERRY

COMMON CHOKECHERRY

COMMON CHOKECHERRY

Quinine-Bush *Purshia stansburyana* (Torr.) Henrickson

ALSO CALLED Mexican cliffrose

SYNONYM *Cowania mexicana* var. *stansburyana* (Torr.) Jepson, *Cowania stansburyana* Torr.

DESCRIPTION Small-leaved evergreen, spreading shrub 3 to 6 feet tall, or sometimes becoming a small scrubby tree to 25 feet in height and 6 inches (15 cm) in trunk diameter, with an irregular open crown of stiff, erect branches.

TWIGS hairy, glandular, and reddish brown the first year, becoming hairless and gray.

LEAVES crowded and small, ¼ to ⅝ inch (6 to 15 mm) long, wedge-shaped, divided into 3 to 7 narrow lobes, thick and leathery with edges rolled under, with white sticky resinous dots, above dark green and loosely hairy or almost hairless, beneath densely white-woolly, bitter.

FLOWERS many but borne singly, large, ¾ to 1 inch (2 to 2.5 cm) across, whitish or pale yellow, fragrant, March to September.

FRUITS 5 to 10 clustered from a flower, ¼ inch (6 mm) long, each with long feathery or plumy, whitish tail (style) 1¼ to 2 inches (3 to 5 cm) long.

BARK shreddy, splitting into long narrow strips, reddish, brown, or gray.

WOOD brown, with thin whitish sapwood.

HABITAT Common on dry rocky hills and plateaus, especially on limestone, in upper desert, desert grassland, oak woodland, and pinyon-juniper woodland zones, 3,500 to 8,000 feet elevation.

NOTE Quinine-Bush is an important browse plant for deer, cattle, and sheep. Covered with whitish flowers, it is beautiful when in full bloom and can be an attractive ornamental in home landscapes. It is also planted for erosion control.

ETYMOLOGY Named for *Frederick Traugott Pursh* (1774-1820), German explorer, collector, horticulturist, author, who made important contributions to America in all his fields during the 21 years he resided there.

QUININE-BUSH

QUININE-BUSH

QUININE-BUSH

Torrey Vauquelinia *Vauquelinia californica* (Torr.) Sarg.

ALSO CALLED Arizona rosewood

DESCRIPTION Evergreen small tree 10 to 30 feet tall, with trunk ½ to 1½ feet in diameter, with stiff twisted branches and slender twigs; or a low, much-branched shrub.

LEAVES lance-shaped, 1½ to 4 inches (3.5 to 10 cm) long and ¼ to ⅝ inch (6 to 15 mm) wide, long-pointed, edges saw-toothed (the teeth often gland-tipped), leathery, bright yellow-green, finely hairy beneath.

FLOWERS many in dense clusters 2 to 3 inches (5 to 7.5 cm) across, small, ⅜ inch (9 mm) across, white, in May and June.

FRUIT a hard capsule ¼ inch (6 mm) long, hairy, splitting into five 2-seeded parts, remaining attached during winter.

BARK thin, broken into small square scales or shaggy, dark red brown.

WOOD hard, very heavy, dark brown streaked with red.

HABITAT Canyons and mountains in upper desert and lower oak woodland, 2,500 to 5,000 feet elevation, mostly in Mexican border region. In New Mexico reported from Guadalupe Canyon (Hidalgo County).

NOTE The hard, heavy wood is beautiful and suggested the name Arizona rosewood. However, commercial use is limited by the small size and scarcity of the trees and their slow growth.

ETYMOLOGY Vauquelinia honors French chemist *Louis Nicholas Vauquelin* (1763-1829).

TORREY VAUQUELINA - FLOWERS

TORREY VAUQUELINA

TORREY VAUQUELINA

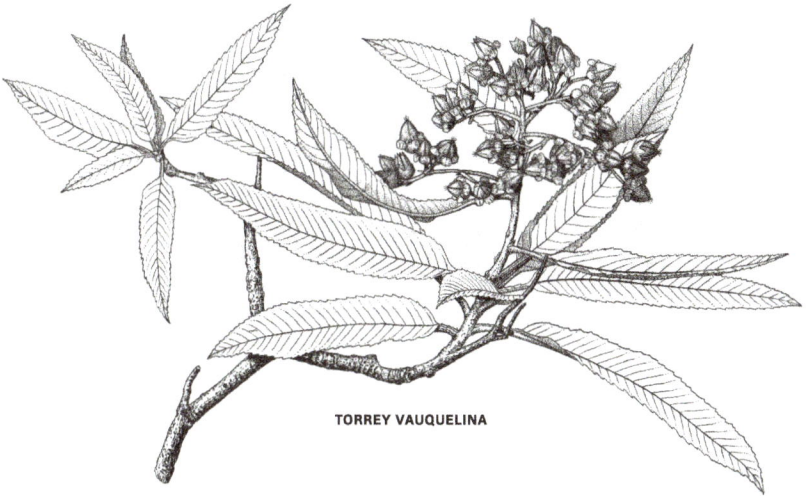

TORREY VAUQUELINA

RUSCACEAE *Butcher's-Broom Family*

Bigelow Nolina *Nolina bigelovii* (Torr.) S. Watson

ALSO CALLED Bigelow's beargrass

DESCRIPTION Narrow-leaved evergreen shrub or rarely small tree 4 to 16 feet tall, resembling yucca, with massive, unbranched trunk 2 to 3 feet in diameter, bearing at the top a dense cluster of stiff, spreading, grasslike leaves and below mostly covered with old dead leaves.

LEAVES numerous, very long and narrow, 2 to 4½ feet (0.6 to 1.4 m) long and ⅜ to ¾ inch (9 to 20 mm) wide, grasslike, stiff and leathery, with edges slightly rough (edges saw-toothed in the variety), separating into long fibers, in age straw-colored and hanging down against the trunk.

FLOWER STALK upright, large, 3 to 8 feet or more in length, the upper half or two-thirds much branched and bearing very many small flowers ⅛ inch (3 mm) or more in length, white with greenish tinge, in June and July.

SEED capsule about ½ inch (12 mm) in diameter, thin and membranous, 3-winged, with few seeds.

HABITAT Rocky hillsides and canyons in desert, 500 to 3,500 feet elevation; not in New Mexico.

ETYMOLOGY Named for *P. C. Nolin,* French agricultural writer c. 1755. This species is named for *John M. Bigelow* (1804-78), surgeon and botanist, who made large plant collections in the Southwest on Government surveys from 1850 to 1854.

ADDITIONAL SPECIES Parry Nolina (*Nolina parryi* S. Watson; synonym *Nolina bigelovii* var. *parryi* (S. Watson) L. Benson), with leaf edges saw-toothed and with slightly larger

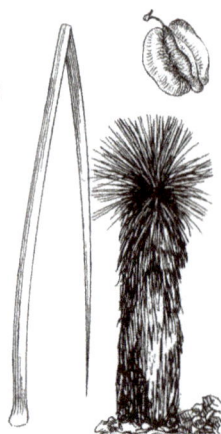

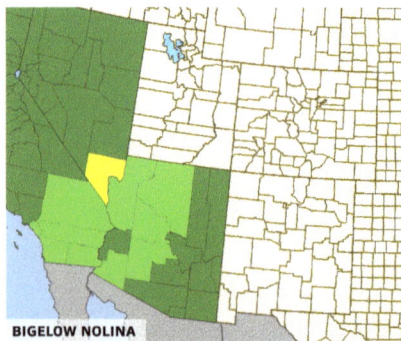

BIGELOW NOLINA

PARRY NOLINA

flowers and fruits than the typical variety, occurs in Arizona as a shrub near Kingman and southward in Mohave County and is the common type in California.

NOTE Bigelow nolina is described in manuals as a shrub and until recent years was not known to reach tree size. However, Leslie N. Goodding discovered small trees of the this species (up to 15 feet in height) in Tinajas Altas Mountains of Yuma County, southwestern Arizona. A number of these oddly shaped desert trees with about the same maximum size are under protection in Joshua Tree National Park, southeastern California. The large flower stalks with beautiful white flower clusters are visible from a long distance as landmarks in the desert.

BIGELOW NOLINA

BIGELOW NOLINA

RUTACEAE *Rue Family*

Ptelea HOPTREE

Shrubs or small trees; leaves commonly trifoliolate, the leaflets lanceolate or ovate, somewhat rhombic; flowers small, perfect or unisexual, in compound cymes; calyx lobes, petals, and stamens 4 or 5; fruit flat, nearly orbicular, winged, mostly 2-celled, indehiscent.

The plants have a strong, somewhat disagreeable odor and are not eaten by livestock. The fruits are reported to have been used as a substitute for hops in beer-brewing. Most of the numerous forms that have been described as species in this genus appear to be only individual variations; there is great diversity in the shape and size of the fruits, as well as of the leaflets. *Ptelea* is a Greek name for an elm tree, but the application here is obscure.

1   Bark of the twigs straw-colored to light olive-colored; leaves yellowish green, paler but not glaucous beneath, often somewhat shiny above, rather thick and firm, glabrous or sparsely pubescent beneath; leaflets prevailingly rhombic-lanceolate but sometimes rhombic-ovate, the terminal one commonly 3 or more times as long as wide . . . . PALE HOPTREE (*Ptelea trifoliata* subsp. *pallida*)
1   Bark of the twigs brown or dark purple (commonly mahogany-colored or plum-colored); leaves bright green or bluish green, often glaucous beneath, not shiny above, thin, glabrate or permanently soft-pubescent beneath; leaflets prevailingly rhombic-ovate but sometimes rhombic-lanceolate, commonly less than 3 times as long as wide . . . . NARROWLEAF HOPTREE (*Ptelea trifoliata* var. *angustifolia*)

Narrowleaf Hoptree *Ptelea trifoliata var. angustifolia* (Benth.) M.E. Jones

SYNONYM *Ptelea angustifolia* Benth.
DESCRIPTION Shrub or small tree to 13 feet or more in height, with strong, unpleasant odor in crushed foliage and twigs.
TWIGS brown or dark purple.
LEAVES compound, long-stalked, with 3 mostly ovate leaflets 1 to 3 inches (2.5 to 7.5 cm) long, long-pointed, edges slightly wavy-toothed or without teeth, thin, with many minute gland dots, green or blue-green but not shiny, beneath often with a bloom and soft-hairy or nearly hairless.

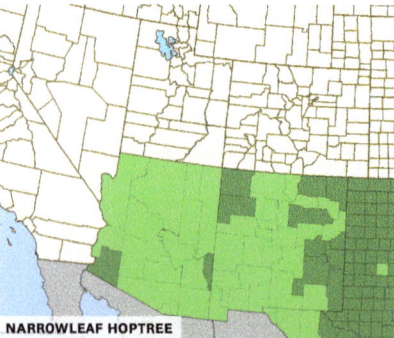

NARROWLEAF HOPTREE

PALE HOPTREE

FLOWERS in branching clusters, partly of separate sexes, small, greenish yellow, from March to June.

FRUIT disk-shaped, flat, light brown, 2- or 3-seeded, surrounded by a broad wing ⅝ inch (15 mm) in diameter.

BARK smooth, brownish gray.

HABITAT Mostly in canyons, ponderosa pine forest and pinyon-juniper woodland, 3,500 to 8,500 feet elevation.

Pale Hoptree *Ptelea trifoliata subsp. pallida* (Greene) V. Bailey

SYNONYM *Ptelea pallida* Greene

DESCRIPTION Shrub or small tree to 10 feet or more in height, with strong, lemon-scented odor of crushed foliage and twigs.

TWIGS straw-colored to greenish yellow.

LEAVES compound, long-stalked, with 3 mostly lance-shaped leaflets 1 to 2 inches (2.5 to 5 cm) long, long-pointed, edges slightly wavy-toothed or without teeth, thick and firm, with many minute gland dots, yellow-green and often shiny above, beneath paler and hairless or sparsely hairy.

FLOWERS in branching clusters, partly of separate sexes, small, greenish yellow, in May.

FRUIT disk-shaped, flat, light brown, 2- or 3-seeded, surrounded by a broad wing ⅝ inch (15 mm) in diameter.

HABITAT Canyons, plateaus, and mountains, pinyon-juniper woodland and upper part of desert zone, 2,000 to 7,000 feet elevation.

NARROWLEAF HOPTREE

NARROWLEAF HOPTREE

NARROWLEAF HOPTREE

SALICACEAE *Willow Family*

Trees or large shrubs, dioecious (the male and female flowers on different plants); leaves deciduous, alternate, simple; flowers of both sexes in catkins, appearing before or with the leaves; ovary 1-celled; stigmas 2 to 4; seeds minute, subtended by silky hairs.

1　Winter buds covered by several scales; scales of the catkins laciniate or fimbriate; flowers borne on broad or cup-shaped disks; stigmas elongate, the lobes slender or dilated. . . . . COTTONWOOD (*Populus*)
1　Winter buds covered by one scale; scales of the catkins entire or merely dentate; flowers without disks; stigmas short. . . . . WILLOW (*Salix*)

Populus COTTONWOOD

Trees, with more or less resinous buds; leaves mostly long-petioled, the blades mostly deltoid or ovate, sometimes lance-shaped; catkins long and drooping; stamens numerous. *Populus,* of obscure derivation, is the classical name of the Poplar.

1　Stigmas 2, 2-lobed, their lobes filiform; leaf-stalks elongated, laterally compressed; buds slightly resinous. Leaves finely serrate; winter-buds glabrous . . QUAKING ASPEN . . . . (*Populus tremuloides*)
1　Stigmas 4, 2-lobed and dilated, their lobes variously divided; buds resinous. 2

2　Leaf-stalks round. 3
2　Leaf-stalks laterally compressed. 4

3　Leaves lanceolate to ovate-lanceolate. . . . . NARROWLEAF COTTONWOOD . . . . (*Populus angustifolia*)
3　Leaves rhombic-lanceolate to ovate . . . . LANCELEAF COTTONWOOD . . . . (*Populus* x*acuminata*)

4　Leaves with glands at tip of the petiole . . . . PLAINS COTTONWOOD . . . . (*Populus deltoides* subsp. *monilifera*)
4　Leaves without glands at tip of the petiole, coarsely serrate, thick. 5

5　Pedicels 2 or 3 times longer than the fruit; leaves broadly deltoid, abruptly short-pointed. . . . . RIO GRANDE COTTONWOOD (*Populus deltoides* subsp. *wislizeni*)
5　Pedicels shorter than the fruit. 6

6　Disk cup-shaped . . . . FREMONT COTTONWOOD (*Populus fremontii*)
6　Disk minute 7

7　Branchlets glabrous; leaves broad-ovate to deltoid, long-pointed and acuminate at apex . . . . PLAINS COTTONWOOD (*Populus deltoides* subsp. *monilifera*)
7　Branchlets pubescent; leaves broad-ovate, abruptly short-pointed or acute at apex . . . . FREMONT COTTONWOOD (*Populus fremontii*)

Narrowleaf Cottonwood *Populus ×angustifolia* James

ALSO CALLED narrowleaf poplar, black cottonwood, mountain cottonwood

DESCRIPTION Medium-sized tree to 50 feet in height and 1½ feet in trunk diameter, with narrow crown and slender, erect branches.

LEAVES lance-shaped, usually 2 to 3 inches (5 to 7.5 cm) long or up to 6 inches (15 cm) on vigorous shoots, narrow and usually less than 1 inch (2.5 cm) wide, tapering and long-pointed, finely saw-toothed, thin but firm, not hairy, bright yellow-green above, paler beneath.

FLOWERS of this genus male and female on different trees (dioecious), in long, narrow catkins, in early spring.

SEED CAPSULES in catkins 2½ to 4 inches (6 to 10 cm) long, with many cottony seeds.

BARK smooth, light yellowgreen, on large trunks becoming fissured and light gray brown.

NARROWLEAF COTTONWOOD

WOOD lightweight, light brown with whitish sapwood.

HABITAT Common to abundant along streams in mountains, growing with willows and alders, ponderosa pine forest and pinyon-juniper woodland zones, 5,000 to 7,000 feet elevation, widely distributed.

NOTE Narrowleaf Cottonwood, easily distinguished by its willowlike leaves, is planted as a shade tree in many places and is suitable for erosion control.

ADDITIONAL SPECIES Lanceleaf cottonwood (*Populus acuminata* Rydb.), formerly thought to be a distinct species, is now considered to be a hybrid between narrowleaf cottonwood and plains cottonwood (*Populus deltoides*). This hybrid differs from narrowleaf cottonwood in its slightly broader, coarsely toothed, ovate leaves with longer leaf-stalks. It has been reported from various parts of New Mexico and Arizona but probably is uncommon in the Southwest and restricted to areas where the two parents meet, such as northeastern New Mexico. Some trees referred to as lanceleaf cottonwood may represent hybrids of narrowleaf cottonwood with other *Populus* species having broad leaves.

LANCELEAF COTTONWOOD

NARROWLEAF COTTONWOOD

NARROWLEAF COTTONWOOD

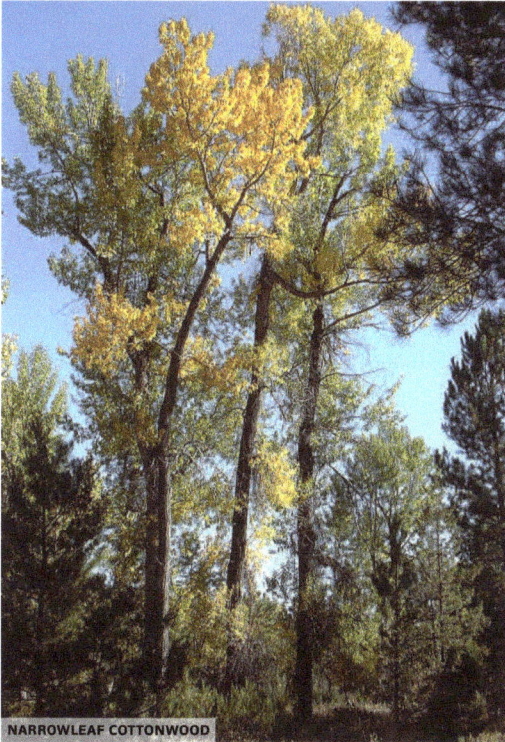

NARROWLEAF COTTONWOOD

Plains Cottonwood *Populus deltoides* subsp. *monilifera* (Aiton) Eckenw.

ALSO CALLED plains poplar

SYNONYM *Populus sargentii* Dode

DESCRIPTION Large tree 60 feet or more in height and 2 to 4 feet or more in trunk diameter, with broad open crown of large branches.

LEAF BLADES ovate, 3 to 4 inches (7.5 to 10 cm) long and broad, long-pointed, coarsely saw-toothed with curved teeth, light green, shiny. Leaf-stalks long, slender, flattened.

SEED CAPSULES in catkins 4 to 8 inches (10 to 20 cm) long, very short-stalked, with many cottony seeds.

BARK thick, deeply furrowed into broad flat ridges, light gray.

WOOD soft, lightweight, brown with whitish sapwood.

HABITAT Along streams in short grass plains, 4,000 to 6,000 feet elevation.

Rio Grande Cottonwood

Populus deltoides subsp. *wislizeni* (S. Watson) Eckenw.

ALSO CALLED Wislizenus cottonwood, valley cottonwood, Rio Grande poplar, alamo

SYNONYM *Populus wislizeni* (S. Wats.) Sarg., *Populus fremontii* var. *wislizeni* S. Wats.

DESCRIPTION Large tree 40 to 100 feet tall and 2 to 4 feet or more in trunk diameter, with widely spreading crown of large branches.

LEAF BLADES broadly triangular, 2 to 2½

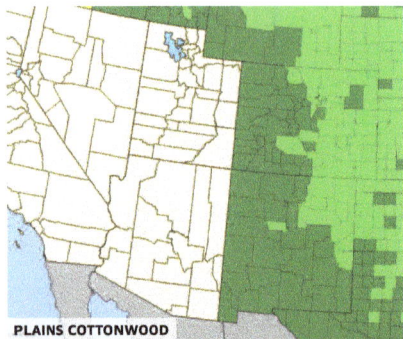

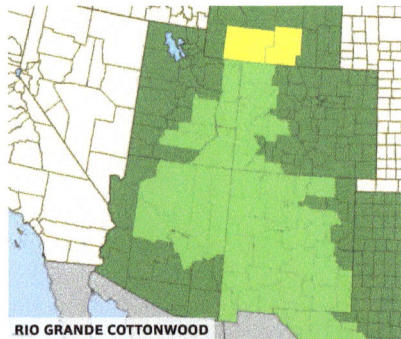

PLAINS COTTONWOOD

RIO GRANDE COTTONWOOD

inches (5 to 6 cm) long and about 3 inches (7.5 cm) wide or sometimes larger, abruptly short-pointed, straight across base, coarsely and irregularly saw-toothed, thick and leathery, shiny yellow green, turning yellow in autumn. Leaf-stalks long, slender, flattened.

SEED CAPSULES in catkins 3½ to 5 inches (8 to 12.5 cm) long, the long stalks ½ to ¾ inch (12 to 20 mm) long and longer than the narrow capsules, with many cottony seeds.

BARK thick, deeply furrowed into broad flat ridges, light gray.

WOOD soft, lightweight, brittle, yellowish brown with whitish sapwood; used for fuel, temporary fence posts, and rafters of buildings..

HABITAT Common along the larger streams in desert, desert grassland, and woodland zones, 3,800 to 6,000 feet elevation. In New Mexico chiefly in Rio Grande drainage northward across the State and in San Juan drainage in northwestern corner, forming extensive woodlands or "bosques" on the broad, sandy floodplains of these two rivers.

NOTE Rio Grande cottonwood, as its name suggests, is the "valley cottonwood" of New Mexico and one of the most common shade trees in the State; widely planted along irrigation ditches.

PLAINS COTTONWOOD - BARK

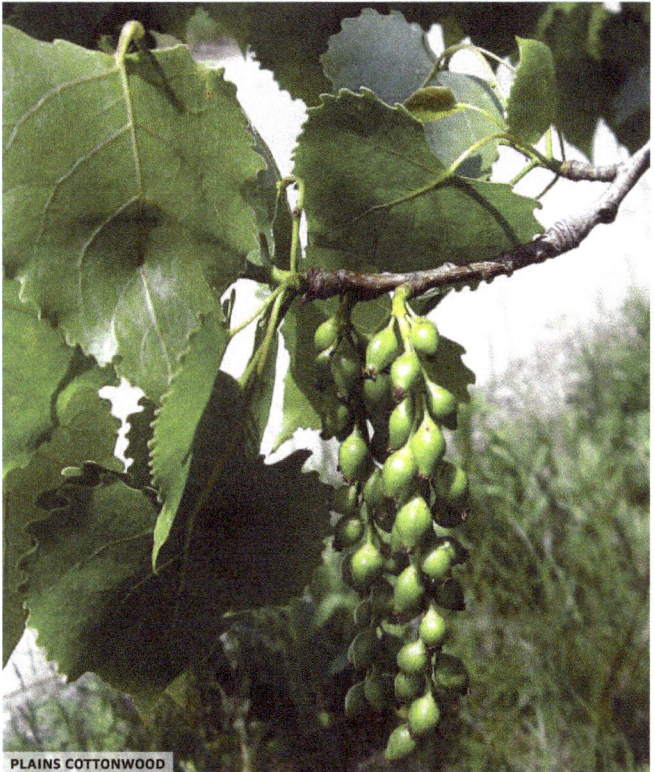

PLAINS COTTONWOOD

Fremont Cottonwood *Populus fremontii* S. Watson

ALSO CALLED Fremont poplar, alamo

SYNONYM *Populus deltoides* subsp. *fremontii* (S. Watson) Kartesz

DESCRIPTION Large tree 50 to 100 feet tall, with trunk 2 to 4 feet or more in diameter, often leaning, and with broad, flat, open crown of large branches.

LEAF BLADES broadly triangular, 2 to 2-¾ inches (5 to 7 cm) long and 2¼ to 3 inches (5.5 to 7.5 cm) wide, short-pointed, nearly straight across base, coarsely and irregularly saw-toothed, thick and leathery, usually not hairy, shiny yellow green, turning yellow in autumn. Leaf-stalks long, slender, flattened.

SEED CAPSULES in catkins 4 to 5 inches (10 to 12.5 cm) long, with the stalks less than ¼ inch (6 mm) long, the egg-shaped capsules ¼ to ½ inch (6 to 12 mm) long, with many cottony seeds.

BARK gray or brown, thick, rough and deeply furrowed.

WOOD light brown with whitish sapwood, soft, lightweight, brittle; used as fuel and for temporary fence posts but not durable.

HABITAT Common along streams and in moist soils of desert, desert grassland, and woodland zones, often associated with Arizona sycamore, willows, and alders, 150 to 6,000 feet elevation.

NOTE This species, the cottonwood of the desert zone in Arizona, is planted extensively as a shade tree and along irrigation ditches.

FREMONT COTTONWOOD

Like other cottonwoods, it is easily propagated from cuttings and is rapidly growing but short-lived. Fremont cottonwood is common at the Grand Canyon, where it is the only large tree in the lower depths and where its green coloring is visible to visitors on the rims above.

FREMONT COTTONWOOD

FREMONT COTTONWOOD

Quaking Aspen *Populus tremuloides* Michx.

ALSO CALLED **aspen, golden aspen, trembling poplar**

DESCRIPTION Small to medium-sized tree usually less than 40 feet in height and 1 foot in trunk diameter, rarely 80 feet tall and 2½ feet in diameter, with slender, rounded crown of thin foliage.

LEAF BLADES nearly round, 1¼ to 3 inches (3 to 7.5 cm) long, short-pointed, rounded at base, finely saw-toothed, shiny green above, dull green beneath, turning golden or orange in autumn before falling. Leaf-stalks 1¼ to 3 inches (3 to 7.5 cm) long, slender and flattened.

SEED CAPSULES in catkins, with many cottony seeds.

BARK smooth, thin, whitish or yellowish, on very large trunks becoming thick, furrowed, and dark gray.

WOOD soft and brittle, lightweight, light brown with very thick whitish sapwood.

HABITAT Common and widely distributed in upper part of ponderosa pine forest, Douglas-fir forest, and spruce-fir forest in high mountains and plateaus, on cool, shaded mountain slopes, in canyons, and along streams, often forming pure stands or groves, 6,500 to 10,000 feet elevation. Range very broad, one of the most extensive of all native North American trees, from Labrador across Canada to Alaska, in northeastern United States south to West Virginia and Missouri, and south in Rocky Mountains and mountains of Pacific coast region to California, Arizona, southwestern Texas, and northern Mexico.

NOTE Quaking aspen is so named because the slender, flattened leaf-stalks enable the leaves to tremble in the slightest breeze. The soft smooth whitish bark makes identification easy.

Aspens are pioneer trees on burned areas, forming thickets of short-lived trees afterwards replaced by conifers. Propagation is chiefly from root sprouts. The wood has been used in the Southwest in manufacture of excelsior and elsewhere also for pulpwood, boxes and crates, food containers, and matches. Domestic livestock and deer browse the foliage within reach.

QUAKING ASPEN

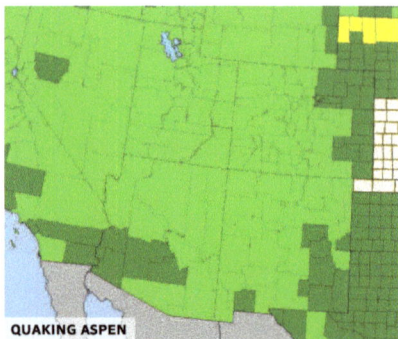

QUAKING ASPEN

QUAKING ASPEN

Salix WILLOW

Shrubs or small trees; leaves from narrowly linear to short-elliptic or obovate; flowers in catkins, appearing before or with the leaves; perianth a single scale; stamens few; pistil single, with a gland at the base of the ovary, the stigma short. Fruit a capsule, containing numerous, very small, hairy seeds. *Salix* is the classical name of the willow tree.

1 Petioles none or very short (not more than ⅛ inch [3 mm] long); leaf blades sericeous, at least beneath, linear-lanceolate or narrowly oblanceolate, seldom more than ¼ inch (6 mm) wide and usually much narrower, the margins entire or remotely denticulate; stamens 2 2

1 Petioles more than ⅛ inch (3 mm) long or, if shorter, the larger blades more than ¼ inch (6 mm) wide, or closely serrate or serrulate 3

2 Leaf blades oblanceolate, sessile or very nearly so, at maturity not more than 1¼ inch (3 cm) long; capsules silky-villous, then glabrate . **YEW-LEAF WILLOW** (*Salix taxifolia*)

2 Leaf blades linear-lanceolate, short-petioled, at maturity 2 inches (5 cm) long or longer; capsules glabrous. . . . . **COYOTE WILLOW** (*Salix exigua*)

3 Margins of the leaf blades entire or nearly so, the lower surface more or less glaucous **4**

3 Margins of the leaf blades closely serrate or serrulate. 7

4 Upper surface of the mature leaves pubescent, not shiny; leaf blades usually less than 3 times as long as wide, elliptic, oblong, or oblong-lanceolate; pistillate catkins seldom more than twice as long as wide at maturity; capsules pubescent . . . . **BEBB'S WILLOW** . . . . (*Salix bebbiana*)

4 Upper surface of the mature leaves glabrous, often shiny 5

5 Leaf blades obovate, rounded to acutish at apex; capsules silky-villous . . . . **SCOULER'S WILLOW** (*Salix scouleriana*)

5 Leaf blades lanceolate or oblanceolate, acute to acuminate at apex; capsules glabrous **6**

6 Leaf blades prevailingly oblanceolate and acute or short-acuminate; stamens 2, the filaments glabrous . . . . **ARROYO WILLOW** (*Salix lasiolepis*)

6 Leaf blades prevailingly lanceolate or oblong-lanceolate and long-acuminate; stamens more than 2, the filaments hairy toward the base . . . . . . **RED WILLOW** (*Salix laevigata*)

7 Lower surface of the leaf blades green, slightly paler than the upper surface but not glaucous. . . . . **GOODDING'S WILLOW** (*Salix gooddingi*)

7 Lower surface of the leaf blades decidedly paler than the upper, usually glaucous 8

8 Leaf blades all, or some of them, conspicuously and sharply long-acuminate at tip; stamens 3 or more; filaments hairy toward the base; usually trees. 9

8 Leaf blades acute or short-acuminate at tip. 12

9 Petioles slender, those of the larger leaves usually ⅜ inch (9 mm) long or longer; blades commonly not more than 3 times as long as wide, not shiny above. . . . . **PEACHLEAF WILLOW** (*Salix amygdaloides*)

9 Petioles stout, usually less than ⅜ inch (9 mm) long; blades commonly at least 4 times as long as wide, often shiny above 10

10 Margins of the leaf blades and the petioles near the apex bearing conspicuous yellowish glands; branchlets and the upper surface of the leaf blades very shiny. . . . . **PACIFIC WILLOW** (*Salix lasiandra*)

10 Margins of the leaf blades and the petioles not or not conspicuously glandular; branchlets and the upper surface of the leaf blades not or only moderately shiny 11

11 Leaf blades commonly broadly lanceolate (less than 6 times as long as wide) , usually

only moderately acuminate, glaucous but ordinarily not silvery white beneath . . . .
. . . . **RED WILLOW** (*Salix laevigata*)

11 Leaf blades commonly narrowly lanceolate (6 or more times as long as wide), very
long- and sharp-acuminate, silvery white beneath . . . . **BONPLAND WILLOW**
. . . . (*Salix bonplandiana*)

12 Leaf blades prevailingly oblanceolate; stamens 2, the filaments glabrous . . . .
. . . . **ARROYO WILLOW** (*Salix lasiolepis*)
12 Leaf blades prevailingly lanceolate or oblong-lanceolate; stamens more than 2, the
filaments hairy toward the base . . . . **RED WILLOW** (*Salix laevigata*)

Peachleaf Willow *Salix amygdaloides* Anderss.

ALSO CALLED **peach willow, almond willow, southwestern peach willow**
SYNONYM *Salix wrightii* Anderss.
DESCRIPTION Small tree to 30 feet tall and 1 foot or more in trunk diameter, with spreading crown.
TWIGS slender, often slightly drooping at tip, yellowish or older twigs gray, hairless.
LEAVES lance-shaped, 2 to 4 (to 7-12) inches (5 to 10 cm) long and ½ to 1¼ (to 1¾) inches (12 to 30 mm) wide, long-pointed at tip, short-pointed or rounded at base, hairless at maturity, yellow green above but whitish to white beneath.
CATKINS 2 to 3 inches (5 to 7.5 cm) long on leafy twigs. Seed capsules long-stalked, hairless, with many cottony seeds.
BARK rough and furrowed, gray or brown.
HABITAT Along streams in short grass, desert, desert grassland, and pinyon-juniper woodland, 3,000 to 7,000 feet elevation.
NOTE Widely distributed from Quebec and New York, west to British Columbia and Washington, and south to Arizona, western Texas, and northern Mexico.

PEACHLEAF WILLOW

Bebb's Willow *Salix bebbiana* Sarg.

ALSO CALLED **beaked willow**

DESCRIPTION Usually a much-branched shrub or a small bushy tree to 15 feet in height, forming clumps or growing singly.

TWIGS slender, branching at wide angles, yellowish to brown, gray-hairy when young but afterward hairless.

LEAVES elliptical, oblong, or reverse lance-shaped, usually small, 1 to 3½ inches (2.5 to 8 cm) long and ⅜ to 1 inch (9 to 25 mm) wide, pointed at both ends or broad at base, edges without teeth or somewhat wavy, thick and firm, dull green above, whitish and roughly net-veined beneath, more or less whitish hairy on both sides but becoming less hairy with age.

CATKINS on short leafy twigs, at maturity 1 to 3 inches (2.5 to 7.5 cm) long and loose. Seed capsules long-stalked, long, very slender, and hairy, with many cottony seeds.

BARK thin and slightly fissured, reddish.

WOOD lightweight, brittle.

HABITAT Moist soils, chiefly along mountain streams in ponderosa pine, Douglas-fir, and spruce-fir forests, 8,500 to 11,000 feet elevation. Widely distributed from Newfoundland and Labrador across Canada to Alaska, south in mountains of western United States to California, Arizona, and New Mexico and south in northeastern States to Nebraska and New Jersey.

NOTE In other regions, where the trees become larger, the wood has been used for baseball bats, charcoal and gunpowder; twigs have been used in making furniture and baskets.

ETYMOLOGY Honors *Michael S. Bebb* (1833-95), American specialist on willows.

BEBB'S WILLOW

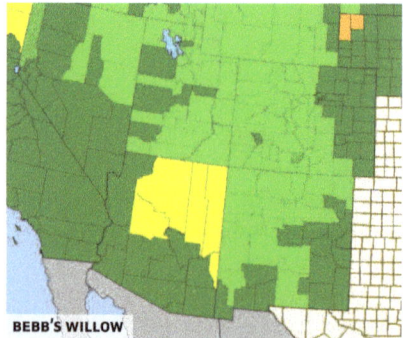

BEBB'S WILLOW

Bonpland Willow *Salix bonplandiana* Kunth

ALSO CALLED **Toumey willow**

DESCRIPTION Small or medium-sized tree 20 to 50 feet in height and 1 to 3 feet in trunk diameter, with broad, rounded crown.

TWIGS slender, red or purple, hairless.

LEAVES narrowly lance-shaped, 4 to 6 inches (10 to 15 cm) long and ½ to ¾ inch (12 to 20 mm) wide, long-pointed, broadest near middle and tapering to base, edges inconspicuously fine-toothed or without teeth, green and shiny above, whitish beneath, without hairs at maturity, shedding irregularly in winter.

CATKINS leafy-stalked, about 1½ inches (3.5 cm) long. Seed capsules reddish yellow, hairless, with many cottony seeds.

BARK rough and fissured or checkered, dark gray or nearly black.

HABITAT Along streams in upper desert, desert grassland, and oak woodland, 2,500 to 5,000 feet elevation, chiefly near Mexican border. Also south through Mexico to Guatemala.

NOTE Bonpland willow is one of the common tree willows in southeastern Arizona.

ETYMOLOGY Named for *Aimee Bonpland* (1773-1858), of France and afterwards South America, who made large botanical collections on an expedition with Alexander von Humboldt from 1799 to 1804 to various Spanish colonies in the New World.

BONPLAND WILLOW

Coyote Willow *Salix exigua* Nutt.

ALSO CALLED **sandbar willow, acequia willow, basket willow**

DESCRIPTION Usually a shrub 6 to 15 feet tall with clustered stems or rarely treelike, forming thickets.

TWIGS yellowish and more or less silveryor gray-hairy.

LEAVES almost stalkless, very narrow, linear or narrowly lance-shaped, 1½ to 3 (or 4) inches (3.5 to 7.5 cm) long and ⅛ to ¼ (or ⅜) inch(3 to 6 mm) wide, short-pointed at both ends, without teeth or with few minute teeth, densely silvery-hairy or silky on both sides, or silvery-hairy while unfolding but soon becoming nearly hairless and yellow-green, or gray-green and more or less densely gray-hairy on both sides (our most typical form).

CATKINS on leafy twigs, 1 to 2 inches (2.5 to 5 cm) long. Seed capsules hairless or nearly so, with many cottony seeds.

BARK smooth or becoming rough and fissured at base of large trunks, gray.

WOOD soft, lightweight, light brown.

HABITAT Common in moist sandy soil along streams in the desert, desert grassland, pinyon-juniper and oak woodlands, and lower ponderosa pine forest, forming dense thickets on sand bars, shores, and washes, from slightly above sea level to 7,000 feet elevation, widely distributed. This species is widely distributed in the Rocky Mountain, Great Basin, and desert regions of the West, from southwestern Texas north to South Dakota and Alberta, west to British Columbia, and south to California, Arizona, and Mexico.

NOTE Coyote willow is the common, thicket-forming shrubby willow in the Southwest; the hairiness (and persistence of the leaf hairs) of the leaves is variable. This species is drought-resistant and especially suitable for planting on stream bottoms to prevent surface erosion. The foliage is excellent browse. The twigs and bark have been used in basket making by Native Americans and others, as the local

COYOTE WILLOW

name basket willow suggests. However, this is not one of the important basket willows of commerce.

ADDITIONAL SPECIES **Sandbar willow** (*Salix interior* Rowlee), a closely related species, is reported from northeastern New Mexico. This species is distinguished from coyote willow by the narrow leaves with edges sharply fine-toothed, silvery-silky on both sides when young but becoming nearly hairless except for hairs remaining on midrib beneath. This wide-ranging species is common in the Midwest and Great Plains.

COYOTE WILLOW

COYOTE WILLOW

Goodding's Willow *Salix gooddingii* Ball

ALSO CALLED **Dudley willow, western black willow**

SYNONYM *Salix nigra* var. *vallieola* Dudley

DESCRIPTION Medium-sized tree 20 to 50 feet in height and 2 to 3 feet in trunk diameter, with broad rounded crown.

TWIGS yellowish, often slightly hairy.

LEAVES narrowly lance-shaped and often slightly curved, 2 to 4 (to 6) inches (5 to 10 cm) long and ¼ to ¾ inches (6 to 20 mm) wide, long-pointed, broadest at base, finely saw-toothed, green or yellowish green on both sides, usually hairless at maturity.

CATKINS 2 to 3½ inches (5 to 8 cm) long on leafy twigs, in March. Seed capsules not crowded, usually hairless, with many cottony seeds.

BARK thick, rough, deeply furrowed with narrow ridges, gray.

HABITAT Common along streams in the desert, desert grassland, and oak woodland, 150 to 5,000 feet elevation.

NOTE Goodding willow, the largest willow in our region, is important for streambank protection because of its deep root system.

ETYMOLOGY Honors *Leslie N. Goodding* (1880-1967), botanist of the United States Department of Agriculture, who made extensive plant collections and many important botanical discoveries in the Southwest.

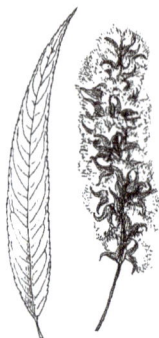

GOODDING'S WILLOW

GOODDING'S WILLOW

Red Willow *Salix laevigata* Bebb

ALSO CALLED polished willow

DESCRIPTION Small or medium-sized tree to 40 feet in height and 2 feet in trunk diameter.

TWIGS yellow to reddish brown.

LEAVES lance-shaped or oblong lance-shaped, 2 to 6 inches (5 to 15 cm) long and ¾ to 1½ inches (2 to 3.5 cm) wide, short-pointed, edges slightly turned under and obscurely saw-toothed, without hairs at maturity, dark green and shiny above, paler beneath.

CATKINS 2 to 4 inches (5 to 10 cm) long on leafy twigs in March. Seed capsules long-stalked, hairless with many cottony seeds.

BARK furrowed into irregular scaly ridges, dark brown.

WOOD soft, lightweight, brittle, light reddish brown with whitish sapwood.

HABITAT Uncommon along streams in oak woodland and pinyon-juniper woodland and sometimes desert zone, 1,800 to 5,000 feet elevation.

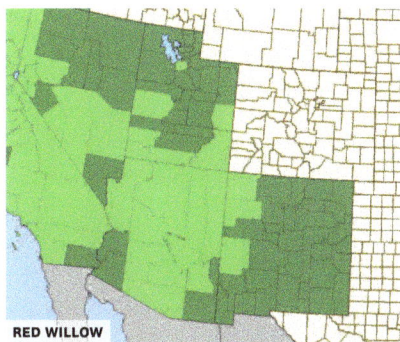

RED WILLOW

Pacific Willow *Salix lasiandra* Benth.

ALSO CALLED **western black willow, yellow willow**

DESCRIPTION Shrub or small tree, rarely to 40 feet or more in height.

TWIGS relatively stout, purple or reddish brown or in early spring bright yellow, hairy when young but becoming hairless and shiny.

LEAVES lance-shaped, 2 to 5 inches (5 to 12.5 cm) long and ½ to 1 inch (12 to 25 mm) wide, long-pointed, mostly rounded at base, edges finely sawtoothed with yellowish gland-tipped teeth, thick and slightly leathery, dark green and very shiny above, paler or whitish beneath, hairless.

FLOWERS of this and other willows male and female on different trees (dioecious), in long, narrow, upright, greenish catkins, in early spring.

CATKINS on leafy stalks, 2 to 4 inches (5 to 10 cm) long at maturity. Seed capsules hairless, with many cottony seeds.

BARK fissured, dark brown.

WOOD brittle, pale brown.

HABITAT Shrubby willow forming clumps along mountain streams in ponderosa pine forest, 5,000 to 7,500 feet elevation.

NOTE A variety of *Salix lasiandra* called **Whiplash willow** (*Salix lasiandra* var. *caudata* (Nutt.) Sudworth) is distinguished by the bright yellow or orange twigs always hairless, and by the very long-pointed, lance-shaped leaves green on both sides. It is a tall bushy shrub which may not reach tree size in the Southwest, where it apparently is uncommon in ponderosa pine forest.

PACIFIC WILLOW

Arroyo Willow　*Salix lasiolepis* Benth.

ALSO CALLED white willow

DESCRIPTION Usually a shrub with clustered stems but sometimes a small tree to 30 feet in height, with slender erect branches forming a narrow, irregular crown.

TWIGS yellow to brown, finely hairy when young.

LEAVES very narrow, linear to narrowly oblanceolate, 2½ to 4 (to 6) inches (6 to 10 cm) long and ⅜ to ¾ inch (9 to 20 mm) wide, short-pointed, edges without teeth or slightly wavy with a few inconspicuous teeth, thick and leathery, dark green above, beneath paler or whitish, becoming nearly hairless.

CATKINS almost stalkless, about 2½ inches (6 cm) long, densely hairy, in March. Seed capsules crowded, dark green, hairless, with many cottony seeds.

BARK smooth, gray brown, on larger trunks becoming rough, fissured into broad ridges, and darker.

WOOD soft, lightweight, brittle, light brown with thick whitish sapwood.

HABITAT Along streams in mountains often with Arizona alder and Arizona sycamore, 6,000 to 7,500 feet elevation.

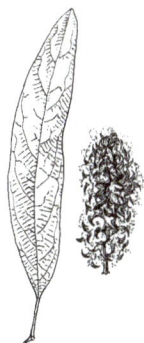

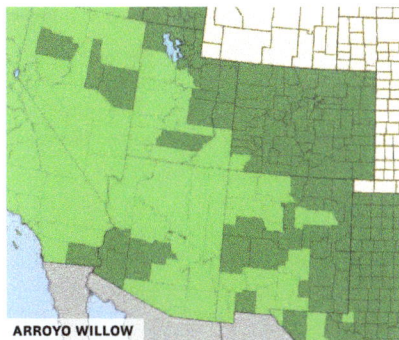

ARROYO WILLOW

Scouler's Willow *Salix scouleriana* Barratt ex Hook.

ALSO CALLED mountain willow, black willow, fire willow

DESCRIPTION Large shrub or rarely a small tree to 4 inches (10 cm) in trunk diameter, with compact rounded crown.

TWIGS stoutish, yellow and densely hairy when young.

LEAVES obovate or elliptical, 1½ to 4 inches (3.5 to 10 cm) long and ½ to 1½ inches (12 to 35 mm) wide, rounded or short-pointed at tip, tapering toward base, edges without teeth or slightly wavy, thick and firm, yellow-green and nearly hairless above, beneath whitish and more or less white or gray-hairy or furry, or becoming rusty hairy in age.

CATKINS stalkless or nearly so, at maturity 1 to 2 inches (2.5 to 5 cm) long, stout and dense. Seed capsules long, slender, gray-woolly, with many cottony seeds.

BARK thin, divided into broad flat ridges, dark brown.

WOOD soft, lightweight, light brown tinged with red and with thick whitish sapwood.

HABITAT Local and uncommon along streams in high mountains, ponderosa pine and Douglas-fir forests, 8,000 to 10,000 feet elevation. Widely distributed in western North America, especially in mountains, from New Mexico north to Black Hills, Montana, Saskatchewan, Yukon, and Alaska, and south to California and Arizona.

NOTE Scouler willow is sometimes called fire willow because it rapidly occupies burned areas in the forests. Unlike most willows, it can grow in the shade of larger trees. This is an important browse species for sheep and cattle where sufficiently common. In the Pacific States it has been planted occasionally for shade and ornament.

TIP The leaves of *Salix scouleriana* are very similar to those of *Salix bebbiana* (both with whitish, net-veined undersurfaces) except for their obovate or oblanceolate leaf blades.

ETYMOLOGY Scouler willow is named for its discoverer, *John Scouler* (1804-71), Scottish naturalist who made a voyage to northwestern America in 1825.

SCOULER'S WILLOW

Yew-Leaf Willow *Salix taxifolia* Kunth

ALSO CALLED yew willow

SYNONYM *Salix exifolia* Dorn, *Salix microphylla* Schlecht. & Cham.

DESCRIPTION Large shrub or small to medium-sized tree 20 to 40 feet tall, with a trunk up to 2 feet in diameter and with a dense, compact, rounded crown.

TWIGS much branched, slender, densely white or silvery-hairy the first year, afterwards gray.

LEAVES very small and densely crowded, very narrow, yewlike, almost stalkless, ½ to 1¼ inches (12 to 30 mm) long and about ⅛ inch (3 mm) wide, short-pointed, the edges usually without teeth, densely silvery-hairy when young and becoming gray-hairy.

CATKINS at the end of short leafy twigs, very short, about ½ inch (12 mm) long, in March and sometimes again in autumn. Seed capsules stalkless, hairy, with many cottony seeds.

BARK rough and fissured, light gray-brown.

HABITAT Infrequent along streams and washes in foothills and mountains, oak woodland and rarely in desert and desert grassland, 3,500 to 6,000 feet elevation, limited to Mexican border region. Also in southwestern Texas and Mexico.

NOTE Yew-leaf willow is a good soil binder, fairly drought-resistant, and excellent browse for livestock but grows slowly. It is an attractive ornamental.

YEW-LEAF WILLOW

SAPINDACEAE *Soapberry Family*

Trees or shrubs ; leaves pinnate or simple ; inflorescences terminal or lateral; flowers perfect or unisexual, small, with or without petals; stamens 4 to 12, commonly 8; fruits berrylike (*Sapindus*) or a winged samara (*Acer*), these united at their base.

1   Fruit a pair of laterally winged samaras, these united near the base; leaves simple and palmately lobed, or palmately divided, or pinnate with few leaflets . . . . . . MAPLE (*Acer*)
1   Fruit leathery and berry-like, not a pair of samaras; leaves pinnately divided . . . .
     . . . . SOAPBERRY (*Sapindus*)

Acer MAPLE

Acer is the Latin name for the maple tree. The word also means sharp and is in reference to the hardness of the wood, which the Romans used for spear hafts.

1   Leaves pinnately compound, 3- to 5-foliolate, the terminal leaflet long-stalked . . . .
     . . . . INLAND BOXELDER (*Acer negundo*)
1   Leaves simple or palmately trifoliolate, the terminal leaflet sessile or short-stalked. . . . **2**

2   Leaf blades thin, glabrous, with numerous acute teeth; inflorescence long-stalked . . . . .
     . . . . ROCKY MOUNTAIN MAPLE (*Acer glabrum*)
2   Leaf blades thickish, usually persistently pubescent beneath, with few obtuse teeth; inflorescence nearly sessile . . . . BIGTOOTH MAPLE (*Acer grandidentatum*)

Rocky Mountain Maple *Acer glabrum* Torr.

ALSO CALLED dwarf maple
SYNONYM *Acer neomexicanum* Greene
DESCRIPTION Shrub or sometimes small tree to 25 feet in height and 1 foot in trunk diameter.
TWIGS reddish brown and without hairs.
LEAVES paired, 3 to 5 inches (7.5 to 12.5 cm) long; leaf-stalks long and often red. Leaf blades 3- or 5-lobed, edges doubly saw-toothed, shiny dark green above and paler beneath, hairless, turning red in fall.
FLOWERS male and female usually on different trees (dioecious), in few-flowered terminal clusters, small, greenish yellow, in May and June.
FRUITS paired, clustered, long-winged "keys," about ¾ inch (2 cm) long, reddish when immature but turning brown.
BARK smooth, thin, gray or brown.
WOOD hard, heavy, light brown.
HABITAT Moist soil and along streams in high mountains, ponderosa pine, Douglas-fir, and spruce-fir forests, 7,000 to 9,000 feet elevation.

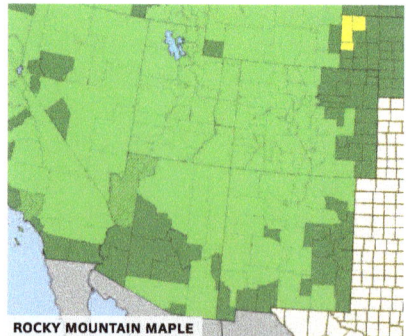

ROCKY MOUNTAIN MAPLE

ROCKY MOUNTAIN MAPLE

Bigtooth Maple   *Acer grandidentatum* Nutt.

ALSO CALLED western sugar maple

DESCRIPTION Small to medium-sized tree to 50 feet tall and 1 foot in trunk diameter, with spreading, rounded crown.

TWIGS bright red and without hairs.

LEAVES paired, long-stalked, heart-shaped, 2 to 5 inches (5 to 12.5 cm) long and broad, 3-lobed, the lobes broad, blunt-pointed, and with few large blunt teeth or small lobes, thick and firm, shiny dark green above, beneath paler and usually finely hairy, turning yellow or red in autumn.

FLOWERS male and female on the same tree in several-flowered clusters, small, about ¼ inch (6 mm) long, yellow, in April.

BIGTOOTH MAPLE

FRUITS paired, clustered, long-winged "keys" 1 inch (2.5 cm) or less in length, greenish.

BARK smoothish or scaly, thin, gray or dark brown.

WOOD hard, heavy, brown or whitish; a good fuel.

HABITAT Fairly common in moist soil in canyons of high mountains and plateaus, ponderosa pine forest, 4,700 to 7,000 feet elevation.

NOTE Bigtooth maple is related to sugar maple of eastern United States and has been used as a source of maple sugar.

BIGTOOTH MAPLE

Inland Boxelder   *Acer negundo* L.

ALSO CALLED **boxelder, Rocky Mountain boxelder**

DESCRIPTION Medium-sized tree to 50 feet in height and 2½ feet in trunk diameter, with broad, rounded crown.

TWIGS greenish, usually finely and densely hairy or hairless and covered with a bloom.

LEAVES paired, long-stalked, pinnately compound, about 6 inches (15 cm) long, with 3 or sometimes 5 leaflets, the end leaflet long-stalked. Leaflets ovate, 2 to 4 inches (5 to 10 cm) long, long-pointed, coarsely saw-toothed, usually thick, slightly hairy beneath or nearly hairless.

FLOWERS male and female on different trees (dioecious), in small several-flowered clusters opening before or with the leaves, very small, yellow-green, in April.

FRUITS paired, clustered, long-winged "keys" 1½ to 2 inches (3.5 to 5 cm) long.

BARK with many narrow fissures and ridges, light gray-brown, on large trunks becoming deeply furrowed.

WOOD soft, lightweight, whitish or pale yellow.

HABITAT Common along streams, chiefly in mountains, oak woodland and ponderosa pine forest zone, 4,000 to 8,000 feet elevation.

NOTE Ours are Inland boxelder (var. *interius* (Britt.) Sarg.), with thicker and more hairy leaves than the typical form of boxelder of the eastern United States. It is a rapidly growing, short-lived tree, often planted for shade where water is available. It escapes from cultivation and has become naturalized locally.

INLAND BOXELDER

INLAND BOXELDER

Western Soapberry *Sapindus saponaria* L.

ALSO CALLED **wild china-tree, cherioni, jaboncillo**

SYNONYM *Sapindus drummondi* Hook. & Arn.

DESCRIPTION Small tree to 25 feet tall and 1 foot in trunk diameter or a large spreading shrub.

TWIGS yellow green, finely hairy.

LEAVES pinnately compound, 5 to 8 inches (12.5 to 20 cm) long, with 13 to 19 leaflets. Leaflets lance-shaped, 1½ to 4 inches (3.5 to 10 cm) long, one-sided, long-pointed, edges not toothed, leathery, pale yellow green, hairless above and slightly hairy beneath.

FLOWERS many in large terminal clusters 6 to 9 inches (15 to 22.5 cm) long, from May to August.

FRUITS yellow, translucent, berrylike, ½ inch (12 mm) in diameter, leathery, 1-seeded, remaining on twigs until spring.

BARK smoothish or fissured and scaly, yellow gray.

WOOD heavy, yellowish.

HABITAT Along streams and canyons, in plains and mountains, plains grassland, upper desert, desert grassland, and oak woodland, 2,400 to 6,000 feet elevation.

NOTE The fruits are poisonous. The foliage is unpalatable and perhaps toxic to livestock. As the common name indicates, the fruits contain relatively high percentages of the alkaloid saponin and have been used for soap in washing clothes.

ETYMOLOGY *Sapo,* soap; the pulp of these trees and shrubs lathers like soap and was used as such by North American indigenous people.

ADDITIONAL SPECIES **Mexican-Buckeye** (*Ungnadia speciosa* Endl.)

This shrub or small tree of southern New Mexico and Texas is characterized by alternate, deciduous, pinnately compound leaves with 5-7 ovate-lanceolate leaflets, each 3-5 inches (7.5 to 12.5 cm) long, thick and dark green; by small, irregular flowers; and by a leathery, 3-valved capsular fruit 2 inches (5 cm) wide, which

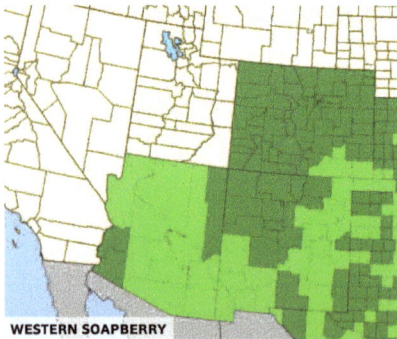

WESTERN SOAPBERRY

MEXICAN-BUCKEYE

contain black, shiny, leathery seeds about ½ inch (12 mm) long, and reputed to be poisonous.

ETYMOLOGY Named for *Baron David von Ungnad,* Austrian Ambassador at Constantinople (1576-1582), who sent horse-chestnut and other seeds to Clusius, a noted Flemish botanist.

MEXICAN-BUCKEYE

MEXICAN-BUCKEYE

WESTERN SOAPBERRY

WESTERN SOAPBERRY

SAPOTACEAE *Sapote Family*

Gum Bumelia *Sideroxylon lanuginosum* Michx.

ALSO CALLED **chittamwood, gum-elastic**
SYNONYM *Bumelia lanuginosa* var. *rigida*
A. Gray, *Bumelia rigida* (A. Gray) Small
DESCRIPTION Spiny shrub or small tree to
13 feet tall, with trunk to 6 inches (15 cm)
in diameter, and with compact crown of
stiff spreading branches.
TWIGS stiff, brown, ending in stout spines,
with an additional spine at base of leaf.
LEAVES single or clustered, elliptical or
reverse lance-shaped, ¾ to 1½ inches (2 to
3.5 cm) long, rounded at tip and narrowed
toward base, edges without teeth, leathery,
with matted white or tan hairs, especially
beneath.
FLOWERS several, in small clusters along
the twigs, small, less than ¼ inch across,
whitish, fragrant, in June.
FRUITS egg-shaped, ⅜ inch (9 mm) or less
in length, purplish black, juicy, 1-seeded.
BARK fissured and scaly, dark gray.
HABITAT Along streams and washes, often
forming thickets, in upper part of desert,
desert grassland, and oak woodland, 3,000
to 5,300 feet elevation, chiefly in Mexican
border region.

The common name refers to the gum
which exudes from cuts in the trunk and
which is chewed by children. Gum bumelia
is the only southwestern representative of
the Sapote Family (Sapotaceae), which is
chiefly tropical. The fruit is acid-sweet, but
after eating it, acid and sour foods like
limes will taste sweet, and the effect can
last for two or three hours.

Sapodilla, the principal tree from which
chewing gum is obtained, is a member of
this family. The highly esteemed tropical
American fruit sapodilla is produced by
Manilkara zapota, a tree that also yields
chicle, from which chewing gum is
manufactured.
ETYMOLOGY *Sideroxylon* is from the Greek
sideros, iron and *xylon,* wood, referring to
the hardness of the heartwood.

GUM BUMELIA

GUM BUMELIA

GUM BUMELIA

GUM BUMELIA - SPINE AT BASE OF LEAF

SIMAROUBACEAE *Ailanthus Family*

Holacantha *Castela emoryi* (A. Gray) Moran & Felger

ALSO CALLED corona de Cristo, crucifixion-thorn

SYNONYM *Holacantha emoryi* A. Gray

DESCRIPTION Very spiny shrub or small tree to 12 feet in height, with numerous stiff, tangled branches and twigs, leafless most of the year.

TWIGS short, stout, ⅛ to ¼ inch (3 to 6 mm) in diameter, finely hairy when young, ending in sharp spines.

LEAVES scalelike, soon shedding.

FLOWERS male and female on different plants (dioecious), in dense many-flowered clusters, small, ¼ to ⅜ inch (6 to 9 mm) in diameter, greenish yellow, in June and July.

FRUIT a ring of 5 to 10 flattened, 1-seeded segments ¼ inch (6 mm) long, the clusters of old fruits remaining attached for years.

BARK smooth.

HABITAT Frequent in desert valleys on clay soils but also on sand dunes, 500 to 2,000 feet elevation; not in New Mexico. Also in southeastern California and in Sonora, Mexico.

NOTE Holacantha (Latin for allthorn) is one of the three very spiny, much branched, shrubby species known in the Southwest as "crucifixion-thorns" (the others are canotia and allthorn). Holacantha is the only native southwestern representative of the Ailanthus Family (Simaroubaceae), which is mostly tropical.

ETYMOLOGY The name honors French naturalist *René Richard Louis Castel.*

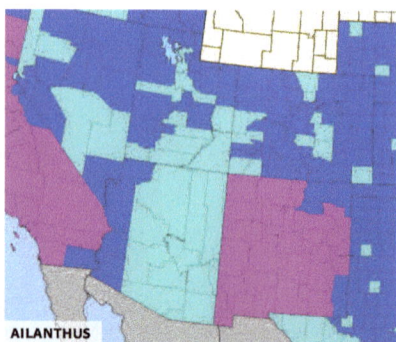

HOLACANTHA

AILANTHUS

ADDITIONAL SPECIES Ailanthus (*Ailanthus altissima* (P. Mill.) Swingle; also called tree-of-heaven and tree-of-heaven ailanthus), a native tree of China belonging to the same family, was formerly planted as a rapidly growing shade tree in southern New Mexico and Arizona. It occasionally escapes from cultivation in the Southwest, spreading by sprouts from the roots, and has become naturalized here (as in many other parts of the United States). These trees have large, pinnately compound leaves, 1½ to 2½ feet (45 to 75 cm) long, with 13 to 25 broadly lance-shaped, long-pointed leaflets 3 to 5 inches (7.5 to 12.5 cm) long, each with 2 to 4 teeth near their base.

HOLACANTHA

HOLACANTHA

AILANTHUS

SOLANACEAE *Nightshade Family*

Tree Tobacco *Nicotiana glauca* Graham

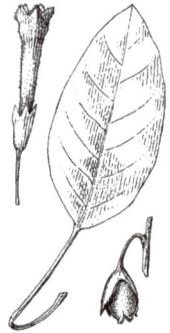

DESCRIPTION Evergreen shrub or small tree 6 to 20 feet tall, with a trunk as much as 3 to 6 inches (7.5 to 15 cm) in diameter.

LEAVES long-stalked, ovate, 2 to 6 inches (5 to 15 cm) long, blunt or short-pointed, edges without teeth, blue-green and covered with a bloom.

FLOWERS in branched clusters at ends of twigs, tubular and narrow, about 1½ inches (3.5 cm) long, yellow, nearly throughout the year.

SEED CAPSULES egg-shaped, ⅜ to ½ inch (9 to 12 mm) long, containing many tiny seeds.

BARK smooth, blue-green.

HABITAT Common and widespread along stream beds, ditches, and washes in desert, from slightly above sea level to 3,000 feet elevation or sometimes higher. Native of Argentina and Chile but extensively naturalized in the tropics.

NOTE Tree tobacco is unusual as an introduced shrub or tree sufficiently hardy to become adapted to moist soils in southwestern deserts. It is planted also as an ornamental and has escaped from cultivation so that, especially along streams, it is a conspicuous feature of local Arizona landscapes.

ETYMOLOGY Named for *Jean Nicot* (1530-1600), French ambassador to Lisbon who introduced tobacco into France.

TREE TOBACCO

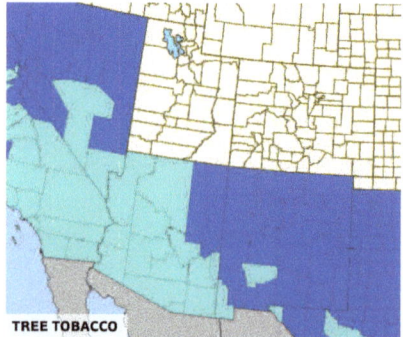

TREE TOBACCO

TREE TOBACCO

TAMARICACEAE *Tamarisk Family*

French Tamarisk *Tamarix gallica* L.

ALSO CALLED salt-cedar

DESCRIPTION Shrub or small tree 10 to 15 feet or more in height, with slender upright or spreading branches and narrow or rounded crown, resembling a juniper though not evergreen.

LEAVES deciduous, many, crowded, scalelike, about ¹⁄₁₆ inch (1.5 mm) long, narrow and pointed, blue-green.

FLOWERS numerous, crowded in many narrow clusters 1 to 2 inches (2.5 to 5 cm) long at ends of twigs, small, ¹⁄₁₆ inch (1.5 mm) long, pink, from March to August.

SEED CAPSULE ⅛ inch (3 mm) long, reddish brown, splitting into 3 to 5 parts, with many tiny hairy seeds.

BARK smoothish, becoming furrowed and ridged, reddish brown.

HABITAT Common or abundant along streams and irrigation ditches, including alkali and salty soils, forming thickets, desert and grassland zones, from slightly above sea level to 5,000 feet elevation; introduced and extensively naturalized, even in the Grand Canyon. Native of

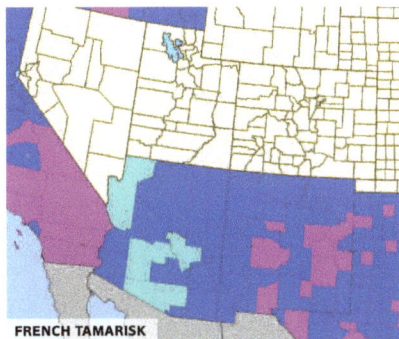

FRENCH TAMARISK

Mediterranean region but escaped and established abundantly in southern and western United States.

French tamarisk, commonly called "salt-cedar" though not related to the cedars or junipers, is sometimes confused with tamarack (*Larix laricina* (Du Roi) K. Koch), an unrelated, coniferous tree of northeastern United States, Canada, and Alaska having a similar name ("tamarack"). Tamarisk is a good erosion-control plant but in places is undesirable. Along Rio Grande and the Salt, Gila, and other rivers, its abundance has created a problem of eradication. It spreads by seeds and is easily propagated by cuttings or by transplanting small wild plants. It grows rapidly and is alkali-tolerant and drought-resistant. The flowers are a source of honey.

ADDITIONAL SPECIES **Athel tamarisk** (*Tamarix aphylla* (L.) H. Karst.), a related species from northeastern Africa and western Asia, is a small to large, fast-growing, evergreen tree planted for shade and windbreaks in southern and central Arizona and to a lesser extent in southern New Mexico. It has escaped from cultivation in a few places and may become naturalized eventually. The wiry, jointed, gray-green twigs are composed of scale leaves 1/16 inch (1.5 mm) long, each encircling the twig and ending in a minute point. The heavy, brittle, light-colored wood is a possible source of furniture and fence posts.

ETYMOLOGY *Tamarix* is the Latin name for these showy deciduous shrubs and trees.

FRENCH TAMARISK

AGAVES & CACTI

ASPARAGACEAE *Asparagus Family*

Yucca YUCCA

Large plants, those included here with a distinct trunk. Leaves numerous, clustered at the ends of the branches, narrow, elongate, and commonly spine-tipped. Flowers large, perfect, numerous in terminal racemes or panicles; perianth segments rather thick, whitish. Fruit dry or fleshy; seeds many, flat. From the Carib name for manihot or cassava, erroneously used here for these evergreen shrubs or small trees having rosettes of sword-shaped leaves.

1   Fruit drying and splitting open at maturity, the opened capsules persistent on the old inflorescence through the winter. SOAPTREE YUCCA (*Yucca elata*)
1   Fruit not splitting open, fleshy or spongy, not persisting on the inflores cence through the winter 2

2   Leaf margin with fine, spreading, effective teeth, fibers not separating from it; tree up to 30 or even 40 feet high, the trunk branching and rebranching repeatedly in old individuals; sepals and petals curved inward, greenish, thick and fleshy; inflorescence 12 to 20 inches (30 to 50 cm) long, a very dense panicle; style none . . . . JOSHUA-TREE . . . . (*Yucca brevifolia*)
2   Leaf margin not toothed, fibers often but not necessarily separating from it; trunk none above ground or the plant with 1 or several sparingly branched stems 2 to 10 feet (0.5-3m) or rarely 20 feet (6 m) high; sepals and petals not curved markedly inward, white or tinged with lavender or purple, not markedly thick and fleshy 3

3   Margins of the leaves not separating into fibers or in age pro ducing a few very fine fibers; branches of the inflorescence densely hairy . . SCHOTT'S YUCCA (*Yucca schottii*)
3   Margins of the leaves eventually separating into fibers; branches of the inflorescence either glabrous or with a few hairs. 4

4   Ovary at flowering time remarkably long and slender, the pistil over all 1 to 3 inches (2.5 to 8 cm) long, the length several times the diameter. . . . . TORREY YUCCA . . . . (*Yucca treculeana*)
4   Ovary at flowering time short and barrel-shaped or short cylindroidal, ⅜ to ½ inch (8 to 12 mm) long . . . . MOHAVE YUCCA (*Yucca schidigera*)

Joshua-Tree *Yucca brevifolia* Engelm.

ALSO CALLED **Joshua-tree yucca**

DESCRIPTION Picturesque or grotesque, narrow-leaved evergreen, small or medium-sized tree 15 to 30 feet (rarely 40 feet or more) in height, with short stout trunk 1 to 3 feet or more in diameter, with open broad crown of many stout, widely spreading or sometimes drooping branches forking at intervals of 2 to 3 feet, bearing at the ends clusters of spreading, grasslike leaves.

JOSHUA-TREE

LEAVES many, long and narrow, 8 to 14 inches (20 to 35 cm) long (or as short as 4 inches (10 cm) in a form), ¼ to ½ inch (6 to 12 mm) wide, stiff, flattened but keeled on outer surface, smooth or slightly rough, blue-green, ending in a short, sharp spine, the edges yellowish and bearing many minute sharp teeth.

FLOWER STALKS at ends of branches, 1 to 1½ feet (30 to 45 cm) long, much branched and bearing flowers nearly to base.

FLOWERS many, crowded in clusters, 1½ to 2½ inches (3.5 to 6 cm) long, greenish yellow, with unpleasant odor, from March to May.

FRUIT egg-shaped, 2½ to 4 inches (6 to 10 cm) long and about 2 inches (5 cm) in diameter, green but becoming brown, fleshy, not splitting open, soon falling.

BRANCHES AND SMALL TRUNKS covered with dead stiff leaves pressed downward; larger trunks corky, rough, deeply furrowed and cracked into plates, brown or gray. Wood lightweight, soft, spongy and pliable, light brown or whitish.

HABITAT The characteristic tree scattered or in "forests," on desert plains in the Mohave desert, 2,000 to 3,500 feet elevation; not in New Mexico.

NOTE Joshua-tree is the outstanding plant of the Mohave desert (located mostly in California and Nevada), and like the saguaro of southern and central Arizona has become a familiar symbol. Ranges of the two monarchs of the desert meet in Arizona at a few places in southern Mohave County. The Mormon pioneers a century ago gave to this species the name of Joshua, a leader of a desert people pointing the way to a new Promised Land, or perhaps praying with uplifted arms.

These weirdly forking plants appear as if survivors of a primitive age. They probably are among the oldest living things in the desert, though not forming annual rings of wood by which age could be counted. Large trees are perhaps 200 to 300 years old, or older. A California giant more than 60 feet tall but no longer living was estimated to be 1,000 years old.

Extensive stands of Joshua-trees are preserved in Joshua Tree National Park in southeastern California. In Arizona, one of the most accessible localities is a large "forest" in southwestern Yavapai County westward along a road about 12 miles west of Congress Junction, on the highway between Phoenix and Prescott.

The spongy, fibrous wood was used in the past for manufacturing paper, for surgeons' splints, wrapping material and boxes, and novelties.

JOSHUA-TREE

JOSHUA-TREE

JOSHUA-TREE

JOSHUA-TREE - FLOWERS

Soaptree Yucca *Yucca elata* (Engelm.) Engelm.

ALSO CALLED palmilla, soapweed

DESCRIPTION Narrow-leaved evergreen, palmlike shrub or small tree, usually 3 to 6 feet tall and infrequently 10 to 15 feet or rarely 25 feet or more in height, with trunk 6 to 9 inches (15 to 22.5 cm) in diameter, unbranched or with few branches, bearing at the top a cluster of spreading grasslike leaves.

LEAVES numerous, very long and narrow, 1 to 2½ feet (30 to 75 cm) long, ⅛ to ⅜ inch (3 to 9 mm) wide, leathery and flexible, yellow green, ending in a sharp spine, with threads along the edges.

FLOWER STALK large, upright in top of plant, 3 to 6 feet (1 to 2 m) or more in length, much branched in upper half, resembling a candelabra.

FLOWERS numerous in large clusters, 1½ to 2 inches (3.5 to 5 cm) long, whitish, in late June and early July.

SEED CAPSULES cylindrical, dry, about 2 inches (5 cm) long, light brown, splitting open in 3 parts to release the many thin black seeds and remaining attached through the winter.

TRUNK bare in lower part, slightly furrowed and gray, in upper part covered by old dead leaves.

WOOD soft and spongy, lightweight, light brown.

HABITAT Common on sandy plains, mesas, and washes, desert grassland and desert, often forming pure stands as the only conspicuous woody species with the grasses, 1,500 to 6,000 feet elevation, widely distributed.

NOTE Soaptree yucca is very common and conspicuous on large areas of desert grassland in southern New Mexico, while related, trunkless species are characteristic of the short-grass plains in other parts of the State. Some very large soaptree yuccas up to 26 feet or more in height are preserved at Jornada Experimental Range in northern Dona Ana County, New Mexico. Yucca is the State flower of New Mexico.

Height growth in soaptree yucca is extremely slow, about 1 inch (2.5 cm) a year. Thus the tallest plants likely range from 200 to 300 years old. When the long, slender or branched trunks lose balance and fall over, they may continue to grow for a time, but sprouts at their bases eventually replace them.

The name soapweed refers to the soapy material in the roots (*amole*) and trunks, which has been used as a substitute for soap. The leaves were employed by the Native Americans in making baskets and are a possible emergency source of coarse fiber. A substitute for jute has been manufactured from the leaf fiber. Native americans reportedly ate the young flower stalks and lower portion of the stem. In drought periods the chopped stems have served as emergency food for cattle. On the range, cattle relish the young flowers.

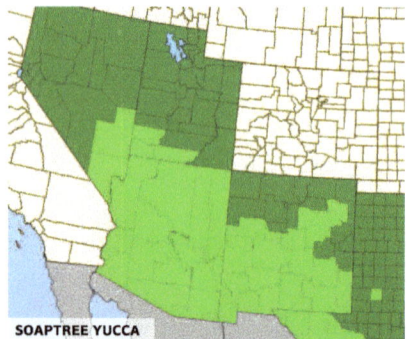

SOAPTREE YUCCA

SOAPTREE YUCCA

Mohave Yucca *Yucca schidigera* Roezl. ex Ortgies

ALSO CALLED Spanish dagger

SYNONYM *Yucca mohavensis* Sarg.

DESCRIPTION Narrow-leaved evergreen shrub or small tree 5 to 16 feet tall and 6 to 10 inches in trunk diameter, usually with few branches or unbranched, bearing at the top a cluster of widely spreading dagger-like leaves.

LEAVES many, very long and narrow, 1½ to 2 feet(45 to 60 cm) (or sometimes 4 feet [1.3 m] or more) in length, ¾ to 1½ inches (2 to 3.5 cm) wide, stiff and stout, concave, broadest at the middle, yellow green, ending in a short, sharp spine, with many coarse threads along the edges.

FLOWER STALK upright, 1½ to 2½ feet (45 to 75 cm) or more in length, projecting only slightly beyond the longest leaves, branched and bearing flowers almost to base.

FLOWERS numerous, crowded in large clusters, 1¼ to 2 inches (3 to 5 cm) long, whitish or cream-colored and often purple-tinged, in March and April.

FRUIT 2 to 3 inches (5 to 7.5 cm) long and 1 to 1½ inches (2.5 to 3.5 cm) in diameter, fleshy, sweetish, and edible, not splitting open, falling before winter.

TRUNK covered with dead leaves or sometimes bare at base, furrowed and gray brown.

WOOD soft and spongy, lightweight, light brown.

HABITAT Rocky slopes and plains in desert, 1,000 to 3,500 feet elevation.

MOJAVE YUCCA

MOJAVE YUCCA

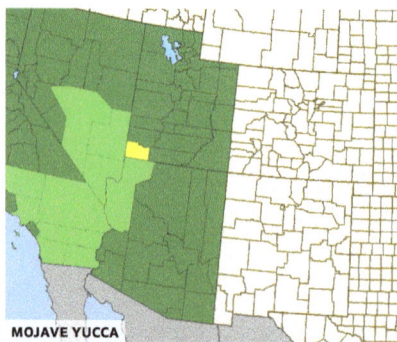

MOJAVE YUCCA

MOJAVE YUCCA

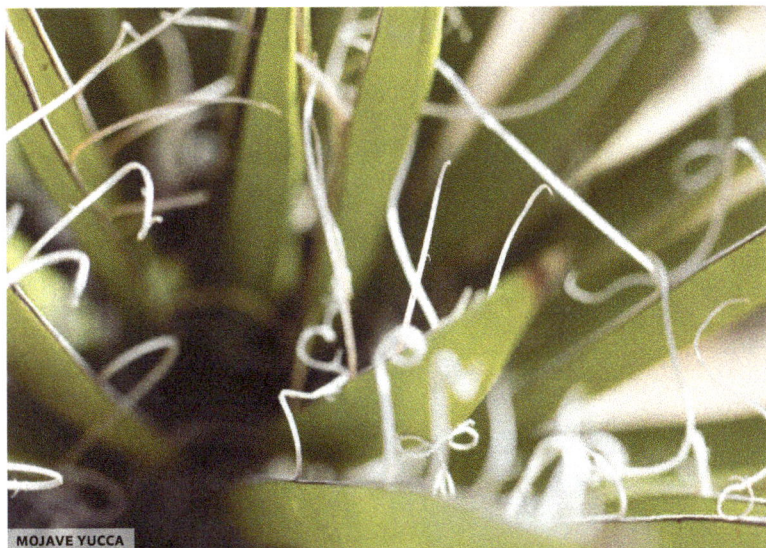

MOJAVE YUCCA

Schott's Yucca *Yucca schottii* Engelm.

ALSO CALLED **mountain yucca, hoary yucca, Spanish bayonet, Spanish dagger**

DESCRIPTION Narrow-leaved evergreen shrub or small tree 5 to 18 feet tall, with usually 2 or 3 trunks from base, the trunks unbranched or with 1 or 2 branches, often leaning slightly, bearing at the top a cluster of spreading bayonetlike leaves.

LEAVES many, very long and narrow, 1½ to 2½ feet (45 to 75 cm) or more in length, 1 to 2 inches (2.5 to 5 cm) wide, flat, leathery and flexible, blue-green, ending in a very short, sharp spine, the edges reddish and without teeth or threads.

FLOWER STALK upright, 1 to 2½ feet (30 to 75 cm) long, finely hairy, much branched and bearing flowers almost to base.

FLOWERS many in large clusters, 1 to 1½ inches (2.5 to 3.5 cm) long, white, from April to August.

FRUIT bananalike, 4 to 5 inches (10 to 12.5 cm) long and 1½ to 2 inches (3.5 to 5 cm) in diameter, green, fleshy, not splitting open, falling before winter.

TRUNK mostly covered with dead leaves, at base rough and scaly, with many horizontal leaf scars, reddish brown or gray.

HABITAT Slopes and canyons in oak woodland and rare in upper desert grassland, extending to a higher elevation than the other tree yuccas, 4,000 to 7,000 feet elevation, mountains chiefly along the Mexican border.

NOTE Schott's yucca is an attractive ornamental in cultivation.

ETYMOLOGY It was discovered by *Arthur C. V. Schott* (1814-75), German-American naturalist, who made large plant collections with the United States-Mexican Boundary Survey in the early 1850's.

SCHOTT'S YUCCA

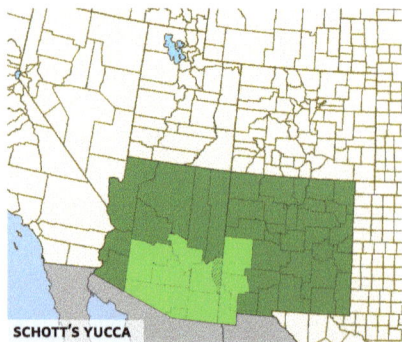

SCHOTT'S YUCCA

Torrey Yucca *Yucca treculeana* Carrière

ALSO CALLED palma, Spanish bayonet, Spanish dagger

SYNONYM *Yucca crassifila* Engelm., *Yucca torreyi* Shafer, *Yucca macrocarpa* (Torr.) Merriam, not Engelm.

DESCRIPTION Narrow-leaved evergreen shrub or small tree 3 to 16 feet tall, with trunk usually single and usually unbranched, 6 to 8 inches (15 to 20 cm) in diameter, bearing at the top a cluster of spreading daggerlike leaves.

LEAVES many, very long and narrow, 2 to 3 feet (0.6 to 1 m) or more in length, 1¼ to 2 inches (3 to 5 cm) wide, stiff and stout, concave, yellow green, ending in a short, sharp spine, with many whitish threads along the edges.

FLOWER STALK upright, 3 to 4 feet (1 to 1.2 m), much branched and bearing flowers nearly to base.

FLOWERS many in large clusters, large, 3 to 4 inches (7.5 to 10 cm) long, bell-shaped, cream-colored, in April or March.

FRUIT banana-like, 4 to 5½ inches (10 to 13 cm) long and 1¼ to 2 inches (3 to 5 cm) in diameter, fleshy, not splitting open, falling before winter.

TRUNK covered with living and dead leaves usually to base, or scaly and dark brown at base.

HABITAT Scattered on mesas, plains, and foothills in desert and desert grassland, 3,700 to 5,000 feet elevation.

NOTE Though Torrey yucca is uncommon, a closely related, trunkless species, **datil yucca** (*Yucca baccata* Torr.) is common and widespread over most of New Mexico and Arizona. The large, fleshy, bananalike fruits of both species are eaten fresh, roasted, or dried by Native Americans, while the leaves were used in basket making. Torrey yucca is grown for ornament and has been used in landscaping along highways.

ETYMOLOGY Honors *John Torrey* (1796-1873), American botanist and chemist of Columbia University, who named many new species of plants of the Southwest.

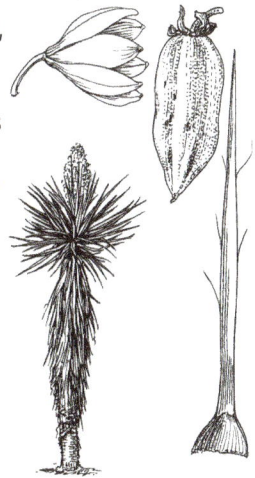

TORREY YUCCA

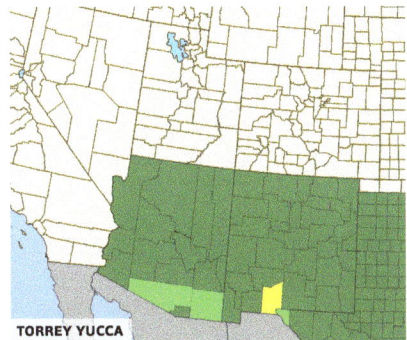

TORREY YUCCA

CACTACEAE *Cactus Family*

Fleshy-stemmed spiny perennials, mostly leafless xerophytes of peculiar aspect; stems globose, cylindric, or flattened, tuberculate or ridged, often jointed, the spines and spicules borne on restricted areas known as areoles. Flowers mostly large and showy; sepals numerous, in several series, gradually becoming petaloid; petals numerous, of delicate texture and handsome colors; stamens very numerous; ovary inferior, with a thick style and several stigmas. Fruit a dry or pulpy berry with thin or thickened rind, and numerous seeds in the single cell.

1   Areoles furnished with glochids (barbed bristles); spines barbed or scabrous . . **CHOLLA** . . . . (*Cylindropuntia*)

1   Areoles not furnished with glochids; spines neither barbed nor scabrous 2

2   Plant with a massive stem more than 12 inches (30 cm) thick, often unbranched but usually bearing smaller, curved lateral branches; flowers white, 4 to 5 inches (10 to 12 cm) long, borne in crownlike clusters at the ends of branches . . . . **SAGUARO** . . . . (*Carnegiea gigantea*)

2   Plant with several to many stems, these of about equal size, less than 8 inches (20 cm) thick, mainly produced from base of the plant; flowers pink 3

3   Spines similar all along the stem; ribs 12 to 17; flowers 2½ to 3 inches (6 to 7.5 cm) long; fruit globose, 2 to 3 inches (5 to 7.5 cm) in diameter, densely spiny . . . . **ORGANPIPE CACTUS** (*Stenocereus thurberi*)

3   Spines twisted and conspicuously longer on the upper (flower-bearing) part of the stem; ribs 5 to 7; flowers 1¼ to 1¾ inches (3 to 4 cm) long, often 2 or more at an areole; fruit globose, ¾ to 1¼ inches (2 to 3 cm) in diameter, unarmed . . . . **SENITA** . . . . (*Pachycereus schottii*)

Saguaro *Carnegiea gigantea* (Engelm.) Britton & Rose

ALSO CALLED giant cactus

SYNONYM *Cereus giganteus* Engelm.

DESCRIPTION Giant columnar tree cactus 20 to 35 feet or rarely 40 feet or more (maximum 52 feet) in height, with a single, continuous, massive, spiny yellow-green trunk 1 to 2 feet or more in diameter, with usually 2 to 10 (or sometimes 20 or more) stout, nearly erect or curved spiny branches.

TRUNK AND BRANCHES cylindrical with rounded or flattened tips, yellow-green, with 12 to 30 prominent vertical fleshy ridges (or ribs) 2 to 4 inches (5 to 10 cm) apart and with alternating grooves, the ridges bearing clusters of about 20 to 25 spreading gray spines to 2 inches (5 cm) or more in length.

LEAVES absent.

FLOWERS many near tops of branches, large and showy, funnel-shaped, 4 to 4½ inches (10 to 12 cm) long and 2 to 3 inches (5 to 7.5 cm) across, with many waxy white petals, with odor like melon, opening at night, in May and June and sometimes again in August.

FRUITS egg-shaped, 2 to 3½ inches (5 to 8.5 cm) long, red, spineless or nearly so,

SAGUARO

fleshy, sweet and edible, splitting open along usually 3 lines and resembling flowers, maturing in June.

WOOD consisting of a framework or ring of vertical rods or ribs separated by long rays and large pith, visible as a skeleton after death and weathering of fleshy parts, the ribs hard, lightweight, light brown.

HABITAT Perhaps the most characteristic tree species of the Arizona desert, common on well-drained rocky or gravelly soils of foothills and slopes and less frequently in valleys, 700 to 3,500 feet (rarely 4,500 feet) elevation; toward the eastern and northern parts of the range confined to south-facing slopes; not in New Mexico; also in Sonora, Mexico.

Saguaro, the State flower of Arizona, has become a symbol for desert landscapes in general. One of the finest stands of mature saguaros is preserved within Saguaro National Park, located about 14 miles east of Tucson. Other good cactus forests are protected in Papago State Park and Phoenix South Mountain Park, both near Phoenix, at Boyce Thompson Southwestern Arboretum near Superior, at Tucson Mountain Park, and at Organ Pipe Cactus National Monument on the Mexican border.

The Pima and Papago gathered the red fruits and eat them fresh, dried, or prepared as preserves, syrup, and beverages; butter was made from the oily seeds. The woody skeleton of ribs is useful in various ways, such as in making shelters, including roof poles and sides of houses, for fences, for kindling fires, and for novelties.

This is the largest cactus in the United States, though other species of similar or greater size occur in Mexico and South America. The champion at Saguaro National Park is 52 feet tall, has 52 arms, an estimated weight of 10 tons, and an age of perhaps 250 years. Increase in height may average 3 inches (7.5 cm) or more a year,

SAGUARO - FLOWERS

varying from 2 to 8 inches (5 to 20 cm), or in the seedling stage only about 1 inch (2.5 cm) annually. Large individuals 30 to 40 feet tall are believed to be as much as 150 to 200 years old and to weigh as much as 6 to 8 tons, most of which is water.

Saguaros are trees exceptionally well-adapted to severe desert conditions, though they thrive in areas with rainfall above average for deserts, locally as much as 15 inches (37.5 cm) a year. Water from rains is absorbed quickly by the shallow roots spreading as much as 50 feet in every direction and is stored in quantity in the thick succulent trunks and branches for growth over the long dry periods. The relatively small surface area exposed, the thick, tough, waterproof outer layer, and absence of leaves all tend to reduce water loss to the hot dry desert air. Specialized ridged green stems without dead bark substitute for leaves in manufacturing food, while the dense spreading spines prevent destruction by desert animals.

SAGUARO

Cylindropuntia **CHOLLA**

Shrubs (ours) with short-jointed stems; joints often tuberculate but never ribbed; leaves small, awl-like; areoles furnished with glochids (barbed bristles); flowers diurnal; fruit indehiscent, often spiny. *Opuntia* (the former genus name) is name of a different plant which grew around Opuntium in ancient Greece.

1   Fruit not persisting for more than 1 season.................... **BUCKHORN CHOLLA**
       .... (*Cylindropuntia acanthocarpa*)
1   Fruit persisting for more than 1 season, fleshy; plants 6 to 13 feet (2 to 4 m) high **2**

2   Mature fruit evidently tuberculate (covered with bumps), solitary; joints ½ to 1¼ inches (1.5 to 3 cm) thick; tubercles ¼ to ⅝ inches (6 to 15 mm) long; spines ⅜ to ⅝ inches (10 to 15 mm) long; flowers purple (occasionally red or yellow)........ **TASAJO**
       .... (*Cylindropuntia spinosior*)
2   Mature fruit slightly or not at all tuberculate **3**

3   Joints readily detached, impenetrably armed, 1¼ to 2 inches (3 to 5 cm) thick, pale green; spines and sheaths straw-colored; fruits proliferous, suspended in chainlike clusters; flowers pink .... **JUMPING CHOLLA** (*Cylindropuntia fulgida*)
3   Joints not readily detached, very spiny but not impenetrably armed, about 1 inch (2.5 cm) thick, elongate, usually purplish; spines dark-colored; fruits solitary or sparingly proliferous; flowers commonly purple, occasionally red or yellow **STAGHORN CHOLLA**
       .... (*Cylindropuntia versicolor*)

Buckhorn Cholla *Cylindropuntia*

acanthocarpa Engelm. & J.M. Bigelow) F.M. Knuth

SYNONYM *Opuntia acanthocarpa* Engelm. & Bigel.
DESCRIPTION Jointed, branching cactus, shrubby or treelike, commonly a shrub 3 to 8 feet tall but rarely a small tree to 15 feet in height, with short trunk and open or dense crown of long spiny branches.
JOINTS OF BRANCHES cylindrical, mostly 6 to 12 inches (15 to 30 cm) long and ¾ to 1 inch (2 to 2.5 cm) or more in diameter, gray-green, fleshy in outer part but becoming woody within, with long narrow tubercles each with a cluster of 6 to 20 brown spines to 1 or 1½ inches (2.5 to 3.5 cm) long (shorter in the variety) in straw-colored sheaths.
FLOWERS large, 1½ to 2½ inches (3.5 to 6 cm) across, usually red or yellow, in April and May.
FRUITS pear-shaped, about 1 inch long, dry and shriveled, spiny, not remaining attached or bearing flowers.
HABITAT Common on sandy flats and washes in desert, 1,500 to 4,000 feet elevation; not in New Mexico.
NOTE This species occasionally becomes treelike in Arizona, though it is usually

BUCKHORN CHOLLA

shrubby. Exceptional individuals of the typical form growing to 15 feet tall are reported in the hills east of Hualpai Mountains (Mohave County) , as are treelike plants of the in the vicinity of Coolidge Dam (Gila County).

BUCKHORN CHOLLA

BUCKHORN CHOLLA

Jumping Cholla *Cylindropuntia fulgida* (Engelm.) F.M. Knuth

ALSO CALLED chainfruit cholla

SYNONYM *Opuntia fulgida* Engelm.

DESCRIPTION Jointed, very spiny, branching cactus, commonly treelike shrub 6 feet or less in height or occasionally a small tree to 15 feet tall. Trunk as much as 6 inches (15 cm) in diameter, usually with several large spreading branches less than 3 feet above the ground forming an irregular open much-branched crown.

JOINTS OF BRANCHES cylindrical, 6 to 8 inches (15 to 20 cm) long, 1¼ to 2 inches (3 to 5 cm) in diameter, pale green but appearing straw-colored because of the numerous spines, fleshy in outer part and becoming woody within, readily detached, bearing many tubercles, each with a cluster of 2 to 12 large brown spines ¾ to 1¼ inch (2 to 3 cm) long and covered with straw-colored sheaths for a year or more; the variety mammillata with thicker branches, larger tubercles and 2 to 6 shorter spines ½ inch (12 mm) long.

LEAVES narrowly cylindrical, ½ to 1 inch (12 to 25 mm) long, light green, fleshy, soon falling.

FLOWERS about 1 inch across, with 5 to 8 petals, pink or white streaked with lavender, from May to August.

FRUITS pear-shaped, 1 to 1⅜ inches (2.5 to 3.5 cm) long and ¾ inch (2 cm) in diameter, green, tubercled but not spiny, fleshy, remaining attached many years and bearing new flowers and fruits annually, the many proliferating fruits hanging down in long branched chains or clusters.

BARK of trunk and larger branches rough, scaly, spineless, black.

WOOD a hollow, perforated, gray cylinder with thick pith and broad rays, hard, lightweight.

HABITAT Often abundant and dominant and forming dense "cactus forests," sandy soils of valleys, plains, and slopes, cactus-shrub desert, 1,000 to 4,500 feet elevation; not in New Mexico.

NOTE Various erect, mostly shrubby cacti with jointed branches are known by the common name cholla (pronounced CHAW-ya or CHO-ya). This name, meaning skull or head in Spanish, may have been suggested by the short, thick branches of some species. Jumping cholla, the largest and perhaps most distinctive of the native chollas, forms dense cactus forests in some areas, especially near Tucson and Florence. Chollas of several species are rapidly spreading and have become a problem on some rangelands in southern Arizona.

JUMPING CHOLLA

JUMPING CHOLLA

Jumping Cholla is easily recognized by the black trunks, the large hanging clusters of proliferating fruits, and the straw color of the joints caused by the numerous spines and sheaths. The impenetrable armed joints are easily detached by touching and adhere to clothing and skin; hence the common name implying that they jump out and attach themselves to passersby. The sharp barbed spines penetrate the flesh, causing pain, and are not easily extracted. Detached branches of chollas will root and start new plants. The dead, weathered wood of old branches, in the shape of a hollow cylinder with numerous holes after decay of the softer tissues, is used in making novelties.

JUMPING CHOLLA

Tasajo *Cylindropuntia spinosior* (Engelm.) F.M. Knuth

ALSO CALLED cane cholla

SYNONYM *Opuntia spinosior* (Engelm.) Toumey

DESCRIPTION Jointed, branching cactus, shrubby or treelike and usually less than 8 feet tall but sometimes a small tree 10 feet tall and 6 inches (15 cm) in trunk diameter, with irregular open crown of many spiny branches. Larger branches usually several at a point and spreading at right angles, smaller branches often 2-forked.

JOINTS OF BRANCHES cylindrical, 4 to 12 inches (10 to 30 cm) long, ¾ to 1¼ inches (2 to 3 cm) in diameter, fleshy in outer part and becoming woody within, bearing very many tubercles, each with a cluster of 10 to 20 usually gray spines ¼ to ½ inch (6 to 12 mm) long.

LEAVES narrowly cylindrical, ⅜ inch (9 mm) long, fleshy, soon falling.

FLOWERS few, clustered at ends of branches, large, 1½ to 2 inches (3.5 to 5 cm) across, with about 10 petals, varying in color from yellow to red, white, or purple, in May and June.

FRUITS pear-shaped, 1 to 1½ inches (2.5 to 3.5 cm) long and broad, bright lemon yellow at maturity, strongly tubercled and becoming spineless, fleshy, remaining attached through the winter. Bark of trunk and larger branches nearly black, ridged and scaly, spineless.

WOOD a hollow, perforated, brown cylinder, hard, lightweight.

HABITAT Common and widely distributed, especially typical of desert grassland in southern Arizona but ranging from desert to oak woodland and pinyon-juniper woodland and occasionally to ponderosa pine forest, 1,200 to 7,000 feet elevation.

NOTE Tasajo (meaning jerked beef in Spanish) is a shrubby cholla in New Mexico but becomes treelike in Arizona. Canes and other novelties are made from the hollow wood of old trunks and large branches, which in this species are long and straight. The fruits are eaten by cattle. Hybrids of this species with jumping cholla, having intermediate characters, have been found along the Gila River near Florence, Pinal County, Arizona.

TASAJO

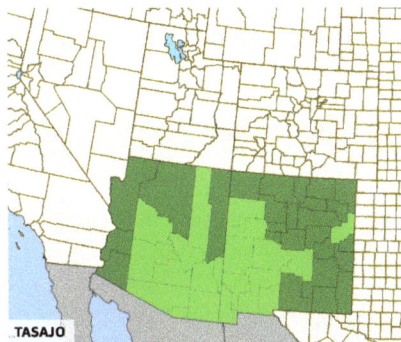

TASAJO

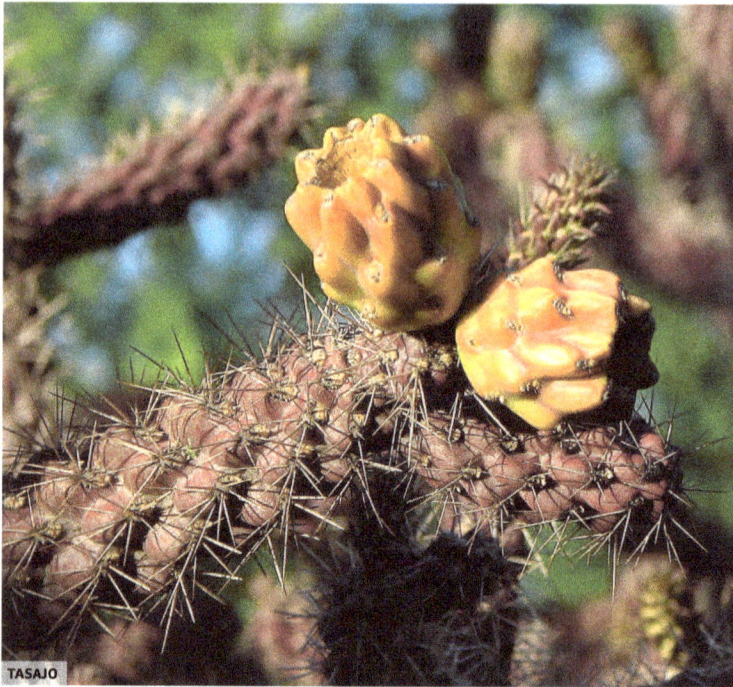

TASAJO

Staghorn Cholla *Cylindropuntia versicolor* (Engelm. ex J.M. Coult.) F.M. Knuth

ALSO CALLED tree cholla, deerhorn cholla

SYNONYM *Opuntia versicolor* Engelm.

DESCRIPTION Jointed, very spiny, branching cactus, treelike shrub or sometimes a small tree up to 12 feet tall, with a trunk as much as 6 inches (15 cm) in diameter and with broad, rounded crown of many spreading branches.

JOINTS OF BRANCHES cylindrical, 4 to 12 inches (10 to 30 cm) or more in length, ⅝ to 1 inch (15 to 25 mm) in diameter, purplish green or dark green, fleshy in outer part and becoming woody within, bearing many long tubercles, each with a cluster of 5 to 10 or more gray or purplish spines ¼ to ⅝ inch (6 to 15 mm) long but of uneven length.

LEAVES narrowly cylindrical, ⅜ inch (9 mm) long, fleshy, soon falling.

FLOWERS few together at ends of branches, large, about 1½ inches (3.5 cm) across, with many petals, varying in color, commonly orange or brown but also yellow, green, or red, in May.

FRUITS single or sometimes 2 or 3 in a chain, pear-shaped, 1 to 1½ inches (2.5 to 3.5 cm) or more in length and ⅝ to 1 inch (15 to 25 mm) in diameter, green tinged with purple or red, tubercled, usually not spiny, fleshy, remaining attached through the winter.

STAGHORN CHOLLA

BARK of trunk light brown or purple, smoothish, becoming scaly at base.
WOOD a hollow, perforated, gray cylinder, hard, lightweight.
HABITAT Valleys and washes, and on watered slopes of foothills and mesas, cactus-shrub desert, 1,200 to 4,000 feet elevation; southern Arizona, not in New Mexico.
NOTE The common name refers to the resemblance of the widely spreading forked branches to a deer's antlers.

STAGHORN CHOLLA

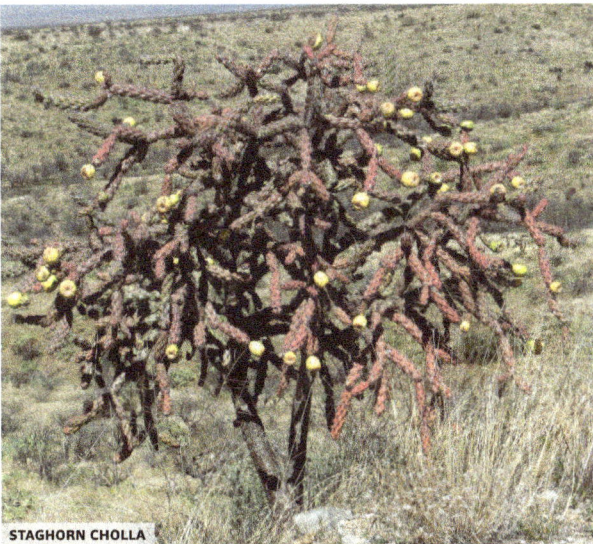

STAGHORN CHOLLA

Senita *Pachycereus schottii* (Engelm.) D.R. Hunt

SYNONYMS *Cereus schottii* Engelm., *Lophocereus schottii* (Engelm.) Britt. & Rose

DESCRIPTION Large treelike cactus 5 to 20 feet tall, without a single trunk, with many (up to 50 or more) erect, columnar, spiny, yellow-green branches from or near the ground and unbranched above.

BRANCHES angular, 4 to 8 inches (10 to 20 cm) in diameter, with usually 5 to 7 prominent vertical fleshy ridges (or ribs) 2 to 4 inches (5 to 10 cm) apart and with alternating grooves, the ridges on lower 4 to 5 feet spineless or with clusters of about 15 or fewer spreading gray spines about ½ inch (12 mm) in length, the ridges on upper part of branch bearing clusters of about 30 gray, twisted, bristlelike spines 1 to 2 inches (2.5 to 5 cm) long.

LEAVES absent.

FLOWERS near tops of branches, 1¼ to 1½ inches (3 to 3.5 cm) long and 1 inch (2.5 cm) across, with pink petals, odorless, opening at night, from April to August.

FRUITS nearly spherical, about 1 inch in diameter, red, almost spineless, fleshy, bursting irregularly.

HABITAT Very rare and local on clay soils of valleys and plains, desert zone, about 1,500 feet elevation, southern Arizona on Mexican border; not in New Mexico. Arizona: in and near southeastern boundary of Organ Pipe Cactus National Monument, western Pima County. Also in Sonora and Lower California, Mexico. Only a small number of these cacti are present in the United States; most are preserved in Organ Pipe Cactus National Monument.

NOTE Senita perhaps may be included as a tree because of its many columnar branches of tree size, though without a definite trunk.

ETYMOLOGY *Pachycereus* is from the Greek *pachys*, thick, and *cereus*, cactus. These cacti have very stout stems. The common name, *senita*, meaning old, or an old person, refers to the conspicuous long bristles, resembling gray hair or beard, in the upper part of the branches.

SENITA

SENITA

SENITA

Organpipe Cactus *Stenocereus thurberi* (Engelm.) Buxb.

ALSO CALLED pitahaya, pitahaya dulce

SYNONYM *Cereus thurberi* Engelm., *Lemaireocereus thurberi* (Engelm.) Britton & Rose

DESCRIPTION Large treelike cactus 10 to 15 feet or up to 25 feet in height, without a single trunk but with many (10 to 20 or more) erect, columnar, spiny, green branches from the ground, unbranched above unless injured.

BRANCHES cylindrical, 5 to 8 inches (12.5 to 20 cm) in diameter, with 12 to 19 prominent vertical fleshy ridges (or ribs) less than 1½ inches (3.5 cm) apart and with alternating grooves, the ridges bearing clusters of about 10 to 19 gray or black spreading spines ½ to 1½ inches (1 to 3.5 cm) long.

LEAVES absent.

FLOWERS scattered near tops of branches, large, 2½ to 3 inches (6 to 7.5 cm) long

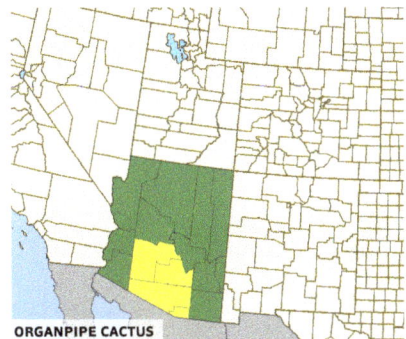

ORGANPIPE CACTUS

and 1½ to 2 inches (3.5 to 5 cm) across, with petals brownish green to greenish white or purple, opening at night, from May to July.

FRUITS egg-shaped or nearly spherical, 2 to 3 inches (5 to 7.5 cm) in diameter, red, with many black spines which are shed at maturity, juicy, sweet and edible, bursting irregularly.

HABITAT Of restricted occurrence on southwestern slopes of mountains in desert, 1,000 to 3,500 feet elevation, in southern Arizona; not in New Mexico.

NOTE Organpipe cactus, with branches slightly suggesting a pipe organ, is of tree size, though without a single trunk, and probably should be included as a tree. This species is protected in the Organ Pipe Cactus National Monument, located on the Mexican boundary south of Ajo in western Pima County. The plants are sensitive to frost, which sometimes kills the growing tips. Papago Native Americans harvest quantities of the sweet fruits.

ORGANPIPE CACTUS

ORGANPIPE CACTUS

PALMS

ARECACEAE *Palm Family*

California Washingtonia *Washingtonia filifera* (Linden) H. Wendl.

ALSO CALLED **California-palm, desertpalm, California fan-palm**

SYNONYM *Washingtonia filamentosa* Kuntze

DESCRIPTION Medium-sized to tall evergreen palm tree 20 to 60 feet or more in height, with thick, columnar, unbranched trunk 2 to 3 feet or more in diameter and with rounded crown of erect and spreading leaves and old dead leaves hanging down against trunk in a thick thatch.

LEAF STALKS 4 to 6 feet (1.2 to 1.8 m) long, stout, with hooked spines along edges, becoming smaller or entirely absent above.

LEAF BLADES fanlike, very large, 6 feet (1.8 m) or more in diameter, gray green, the outer part split into many narrow, folded, leathery segments with edges frayed into many threadlike fibers.

FLOWER CLUSTERS large and branched, 10 to 12 feet (3.3 to 3.7 m) long, drooping, bearing numerous small flowers ⅜ inch (9 mm) long, white, slightly fragrant.

BERRY nearly ½ inch (12 mm) long, black, with thin edible flesh and 1 large seed.

TRUNK rough and checkered, with vertical cracks more prominent than the horizontal lines, grayish brown, in upper part covered by mass of dead leaves.

WOOD without annual rings, soft and lightweight, yellowish.

HABITAT Rare and local in canyons of desert mountains, best known from Kofa Mountains in southwestern Arizona, at about 2,500 feet elevation; about 100 wild trees in Palm Canyon and other deep, narrow canyons of Kofa Mountains, about 25 miles southeast of Quartzsite, Yuma County. Also in southeastern California, such as Palm Canyon near Palm Springs and Thousand Palms Canyon near Indio, and in northern Lower California, Mexico.

NOTE California Washingtonias are commonly planted as ornamentals in southern Arizona but were not known to be native in the State until discovered in the Kofa Mountains in 1923. In southeastern California, where this species is more common, the fruits were gathered and eaten by Native Americans.

ADDITIONAL SPECIES A related species from Lower California, **Mexican Washingtonia** (*W. robusta* H. Wendl.), distinguished by more slender and taller trunks about 1 foot (30 cm) in diameter

CALIFORNIA WASHINGTONIA

above the wider base, is also frequent in cultivation. Both species are widely planted in subtropical portions of the United States and in other parts of the world.

ETYMOLOGY Washingtonia is dedicated to President George Washington (1732-99).

KEY FEATURES Washingtonia is distinguished from the commonly cultivated date palms by the arrangement of the leaf blades; in Washingtonia the blade is nearly circular in outline, and the divisions extend only half or two thirds of the way to the base, being arranged palmately (like the fingers of a hand). In date palms the blade is greatly elongated and divided all the way to the midrib, with the separate leaflets pinnately arranged (like the segments of a feather) along the axis.

CALIFORNIA WASHINGTONIA

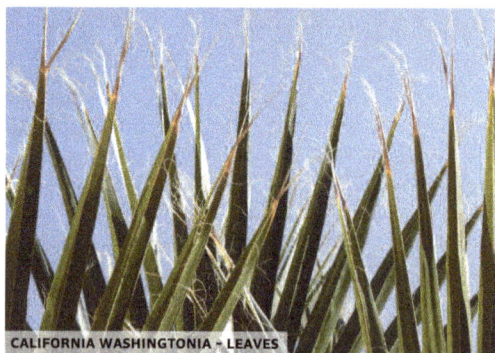

CALIFORNIA WASHINGTONIA - LEAVES

Glossary

Abortive. Imperfectly or not developed; barren.

Acicular. Slenderly needle-shaped.

Accrescent. Increasing in size with age.

Achene. A dry indehiscent, 1-celled and 1-seeded fruit or carpel.

Acuminate. Gradually tapering to the apex; long-pointed.

Acute. Sharply pointed, but not drawn out.

Adnate. Descriptive of unlike organs or parts fused together.

Alternate. Scattered singly along axis; not opposite.

Ament. A scaly, bracted spike of usually unisexual flowers, frequently deciduous in one piece.

Angiosperms. Plants with seeds borne in an ovary.

Anther. The pollen-bearing part of the stamen.

Anthesis. The time when fertilization takes place or a flower expands.

Apetalous. Without petals.

Apiculate. Ending in a minute, short, pointed tip.

Apophysis. That part of a cone scale which is exposed when the cone is closed.

Appressed. Lying close and flat against.

Arborescent. Attaining the size or character of a tree.

Aril. An appendage growing out from the hilum and covering the seed partly or wholly.

Attenuate. Slenderly tapering; acuminate.

Awl-shaped. Tapering from the base to a slender and stiff point.

Axil. The upper angle formed by a leaf or branch with the stem.

Axillary. Situated in an axil.

Baccate. Berry-like; pulpy throughout.

Berry. A fleshy or pulpy fruit with immersed seeds.

Blade. The expanded portion of a leaf.

Bloom. A powdery or waxy substance easily rubbed off.

Bole. The stem of a tree.

Boss. A raised projection, usually pointed.

Bract. A modified leaf subtending a flower or belonging to an inflorescence.

Bractlet. The bract of a pedicel or ultimate flower stalk; a secondary bract.

Bud. The undeveloped state of a branch or flower cluster, with or without scales.

Bud scales. Modified leaves covering a bud.

Bundle (leaf). Strand of fibro-vascular tissue found in cross-section of leaf.

Caducous. Falling off very early.

Calyx. The flower-cup or exterior part of a perianth.

Campanulate. Bell-shaped.

Canescent. Gray-pubescent and hoary.

Capsule. A dry fruit of more than one carpel which splits at maturity to release its seeds.

Carpel. A simple pistil or an element of a compound pistil.

Catkin. The same as an ament.

Caudate. Furnished with a tail or with a slender tip.

Cell. The unit of structure of living things; a cavity of an ovary or anther.

Chambered. Said of pith which is interrupted by hollow spaces.

Ciliate. Fringed with hairs on the margin.

Cone. A fruit with woody, overlapping scales.

Coniferous. Pertains to cone-bearing trees.

Coppice. Growth arising from sprouts at the stump.

Cordate. Heart-shaped.

Coriaceous. Of the texture of leather.

Corolla. Inner part of the perianth, composed of petals.

Corymb. A flat-topped flower cluster, the flowers opening from the outside inward.

Crenate. Dentate with the teeth much rounded.

Crenulate. Diminutive of crenate, finely crenate.

Crown. The upper part of a tree, including the living branches with their foliage.

Cuneate. Wedge-shaped, or triangular with an acute angle downward.

Cuspidate. Tipped with a sharp, rigid point.

Cylindric. Shaped like a cylinder.

Cyme. A flat-topped flower cluster, the flowers opening from the center outward.

Deciduous. Not persistent; falling away as the leaves of a tree in autumn.

Decurrent. Running down, as of the blades of leaves extending down their petioles.

Decussate. In pairs alternately crossing at right angles.

Dehiscent. The opening of an anther or capsule by slits or valves.

Deltoid. Delta-shaped, triangular.

Dentate. Toothed, with the teeth directed outward.

Denticulate. Minutely toothed.

Diadelphous. Stamens formed into two groups through the union of their filaments.

Dimorphous. Occurring in two forms.

Dioecious. Unisexual, the staminate and pistillate flowers on different individuals.

Dissemination. The spreading abroad of ripe seeds from the parent plant.

Divergent. Spreading apart; pointing away.

Dorsal. Relating to the back or outer surface of an organ; the lower surface of a leaf.

Downy. Clothed with a coat of soft, fine hairs.

Drupaceous. Resembling or relating to a drupe.

Drupe. A stone fruit, such as a plum.

E. A latin prefix denoting that parts are missing, as eglandular, without glands.

Ellipsoidal. Of the shape of an elliptical solid.

Elliptic. Of the form of an elipse.

Emarginate. Notched at the apex.

Entire. Leaf margin without divisions, lobes or teeth.

Erose. Descriptive of an irregularly toothed or eroded margin.

Excrescences. Warty outgrowths or protuberances.

Exfoliate. To cleave or peel off in thin layers.

Exserted. Prolonged beyond the surrounding organs, as stamens from the corolla.

Falcate. Scythe- or sikle-shaped.

Fascicle. Dense cluster or bundle.

Fibro-vascular. Consisting of woody fibers and ducts.

Filament. The stalk of an anther.

Fluted. Regularly marked by alternating ridges and groovelike depressions.

Foliaceous. Leaflike in texture or appearance.

Fugacious. Falling or withering away very early.

Fulvous. Tawny; dull yellow with gray.

Furrowed. With longitudinal channels or grooves.

Gibbous. Swollen on one side.

Glabrous. Smooth, not pubescent or hairy.

Gland. Secreting surface or structure; a protuberance having appearance of such an organ.

Glandular. Furnished with glands.
Glaucous. Covered or whitened with a bloom.
Globose. Spherical in form or nearly so.
Gymosperms. Plants with naked seeds; i.e., not enclosed in an ovary.

Habit. The general appearance of a plant; best seen from a distance.
Habitat. The place where a plant naturally grows.
Hilum. The scar or place of attachment of a seed.
Hirsute. Covered with rather coarse or stiff, long hairs.
Hispid. With rigid or bristly hairs.
Hoary. Covered with a close, whitish or gray-white pubescence.
Hybrid. A cross between two nearly related species.

Imbricate. Overlapping, like shingles on a roof.
Indehiscent. Not splitting open; remaining closed.
Inferior ovary. Appearing to grow below the adnate calyx.
Inserted. Attached to or growing out of.
Intolerant. Not capable of doing well under dense forest cover.
Involucre. A circle of bracts surrounding a flower cluster.
Irregular flower. Not symmetrical, similar parts of different shapes or sizes.

Keeled. With a central ridge like the keel of a boat.

Laciniate. Cut into narrow, pointed lobes.
Lanceolate. Lance-shaped.
Lateral. Situated on the side; not at apex.
Leaflet. One of the small blades of a compound leaf.
Leaf scar. Scar left on twig by the falling of a leaf.

Legume. Fruit of the pea family; podlike and splitting open by both sutures.
Lenticel. Corky growth on young bark which admits air to the interior of a twig or branch.
Linear. Long and narrow, with parallel edges.
Lobe. A somewhat rounded division of an organ.
Lobulate. Divided into small lobes.
Lustrous. Glossy, shining.

Membranaceous. Thin and somewhat translucent.
Midrib. The central vein of a leaf or leaflet.
Monoecious. The stamens and pistils in separate flowers but borne on the same individual.
Mucro. A small and abrupt tip to a leaf.
Mucronate. Furnished with a mucro.

Naked buds. Buds without scales.
Nut. A hard and indehiscent, 1-seeded pericarp produced from a compound ovary.
Nutlet. A diminutive nut or stone.

Ob. Latin prefix signifying inversion.
Obconic. Inverted cone-shaped.
Oblanceolate. Lanceolate, with the broadest part toward the tip.
Oblong. About three times longer than broad with nearly parallel sides.
Oblique. Slanting or with unequal sides.
Obcordate. Inverted heart shape.
Obovate. Ovate with the broader end toward the apex.
Obovoid. An ovate solid with the broadest part toward the tip.
Obtuse. Blunt or rounded at apex.
Odd-pinnate leaf. Pinnate with a terminal leaflet.
Orbicular. A flat body circular in outline.
Oval. Broad elliptic, rounded at ends and about 1-1/2 times longer than wide.
Ovary. The part of a pistil that contains the ovules.

Ovate. Shaped like the longitudinal section of an egg, with the broad end basal.

Ovoid. Solid ovate or solid oval.

Ovule. The part of the flower which after fertilization becomes the seed.

Palmate. Radiately lobed or divided, veins arising from one point.

Panicle. A loose, compound flower cluster.

Papilionaceous. Butterfly-like; typical flower shape of legumes.

Pedicel. Stalk of a single flower in a compound inflorescence.

Pedicellate. Borne on a pedicel.

Peduncle. A general flower stalk supporting either a cluster of flowers or a solitary flower.

Peltate. Shield-shaped and attached by its lower surface to the central stalk.

Pendent. Hanging downward.

Pendulous. More or less hanging or declined.

Perfect. Flower with both stamens and pistil.

Perianth. The calyx and corolla of a flower considered as a whole.

Persistent. Remaining attached, not falling off.

Petiolate. Having a petiole.

Petiole. The footstalk of a leaf.

Petiolule. Footstalk of a leaflet.

Pilose. Hairy, with soft and distinct hairs.

Pinnate. A compound leaf with leaflets arranged along each side of a common petiole.

Pistil. Female organ of a flower, consisting of ovary, style, and stigma.

Pistillate. Female flowers without fertile stamens.

Pith. The central, softer part of a stem.

Pollen. The fecundating grains borne in the anther.

Polygamo-dioecious. Flowers sometimes perfect, sometimes unisexual and dioecious.

Polygamo-monoecious. Flowers sometimes perfect and sometimes unisexual, the 2 forms borne on the same individual.

Polygamous. Flowers sometimes perfect and sometimes unisexual.

Pome. An inferior fruit of 2 or several carpels enclosed in thick flesh; an apple.

Prickle. A small spine growing from the bark.

Prostrate. Lying flat on the ground.

Puberulous. Minutely pubescent.

Pubescent. Clothed with soft, short hairs.

Pungent. Terminating in a rigid, sharp point; acrid.

Pyramidal. Shaped like a pyramid.

Raceme. A simple inflorescence of stalked flowers on a more or less elongated rachis.

Racemose. In racemes; resembling racemes.

Rachis. An axis bearing flowers or leaflets.

Receptacle. The more or less expanded portion of an axis which bears the organs of a flower or the collected flowers of a head.

Recurved. Curving downward or backward.

Reflexed. Abruptly turned downward.

Remotely. Scattered, not close together.

Reniform. Kidney-shaped.

Repand. With a slightly sinuate margin.

Reticulate. Netted.

Retrorsely. Directed backward or downward.

Revolute. Rolled backward, margin rolled toward the lower side.

Rhombic. Having the shape of a rhombus.

Rufous. Red-brown.

Rugose. Wrinkled.

Samara. An indehiscent, winged fruit, as in maple and ash.

Scabrous. Rough to the touch.

Scarious. Thin, dry, membranaceous, not green.

Scorpioid. A form of unilateral inflorescence circinately coiled in the bud.

Scurfy. Covered with small branlike scales.

Serrate. Toothed, the teeth pointing upward or forward.

Sessile. Without a stalk.

Sheath. A tubular envelope, or enrolled part or organ.

Shrub. A woody, bushy plant, branched at or near the base and usually less than 15 feet in height.

Sinuate. With a strong, wavy margin.

Sinus. The cleft or space between two lobes.

Spike. A simple inflorescence of sessile flowers arranged on a common, elongated axis.

Spine. A sharp, woody outgrowth from a stem.

Spinescent. With short, rigid branches resembling spines.

Spinose. Furnished with spines.

Stamen. The pollen-bearing organ of the male flower.

Staminate. Male flowers provided with stamens but without pistils.

Stellate. Star-shaped.

Sterigmata. Short, persistent leaf bases found on spruces and hemlocks.

Stigma. The part or surface of a pistil which receives pollen for the fecundation of the ovules.

Stipule. An appendage at the base of the petiole, usually one on each side.

Stoma. An orifice in the epidermis of a leaf used to connect internal cavities with air.

Stomata. Plural of stoma.

Stomatiferous. Furnished with stomata.

Strobile. A cone.

Style. The attenuated portion of a pistil between the ovary and the stigma.

Sub. A Latin prefix denoting somewhat or slightly.

Subtend. To lie under or opposite to.

Subulate. Awl-shaped.

Succulent. Juicy; fleshy.

Superior ovary. Free from and inserted above calyx; hypogynous.

Suture. A junction or line of dehiscence.

Taproot. The primary descending root, which may be either very large or absent at the maturity of the tree.

Terete. Circular in traverse section.

Terminal. Situated at the end of a branch.

Ternate. In groups of three.

Tolerant. Capable of enduring shade.

Tomentose. Densely pubescent with matted wool or tomentum.

Torulose. Cylindric, with swollen partitions at intervals.

Tree. A plant with a woody stem, unbranched at or near base, and at least 8 feet in height and 2 inches in diameter.

Truncate. Ending abruptly, as if cut off at the end.

Tubercle. A small tuber or excrescence.

Turbinate. Top-shaped.

Undulate. With wavy surface or margin.

Unisexual. Of one sex, either staminate or pistillate.

Valvate. Leaf buds meeting at the edges, not overlapping.

Veins. Threads of fibro-vascular tissue in a leaf or other flat organ.

Ventral. Belonging to the anterior or inner face of an organ; the upper surface of a leaf.

Vernal. Appearing in the spring.

Villose. Hairy with long and soft hairs.

Whorled. Three or more organs arranged in a circle round an axis.

Wing. A membranous or thin and dry expansion or appendage of an organ.

Woolly. Covered with long and matted or tangled hairs.

Ecoregions of Arizona

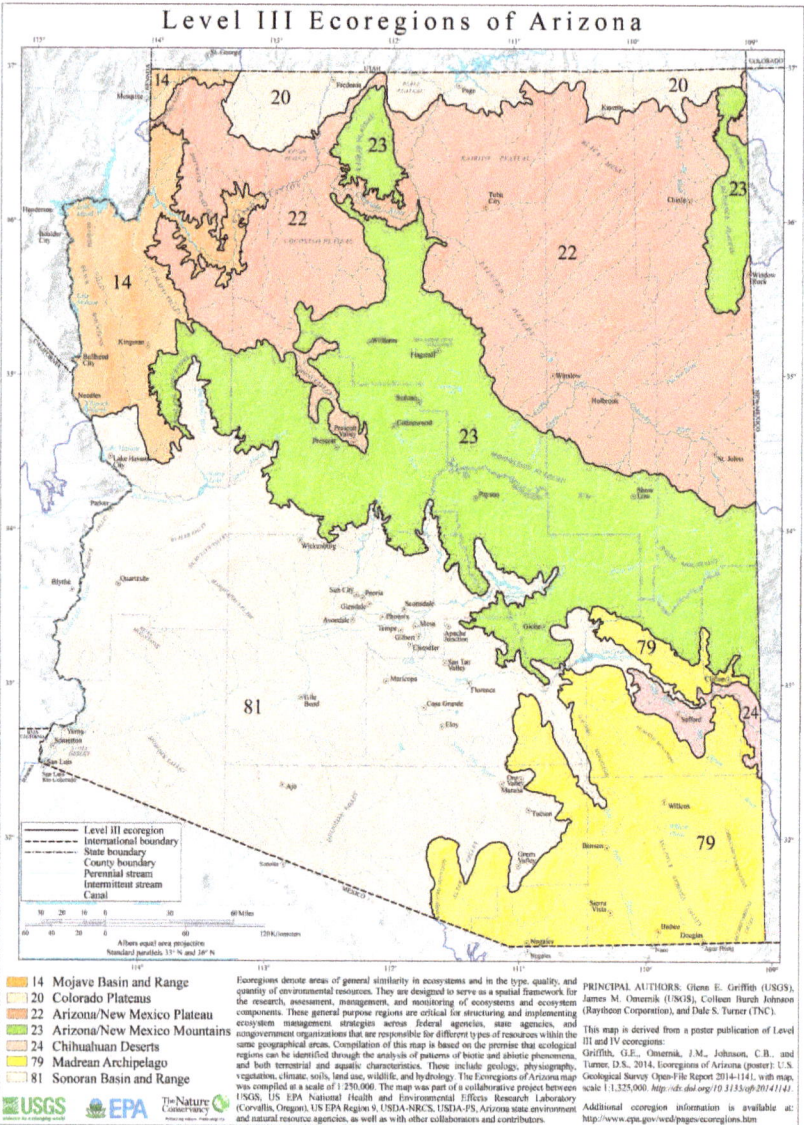

Level III Ecoregions of Arizona

14 Mojave Basin and Range
20 Colorado Plateaus
22 Arizona/New Mexico Plateau
23 Arizona/New Mexico Mountains
24 Chihuahuan Deserts
79 Madrean Archipelago
81 Sonoran Basin and Range

Ecoregions of New Mexico

Level III Ecoregions of New Mexico

- 20 Colorado Plateaus
- 21 Southern Rockies
- 22 Arizona/New Mexico Plateau
- 23 Arizona/New Mexico Mountains
- 24 Chihuahuan Deserts
- 25 High Plains
- 26 Southwestern Tablelands
- 79 Madrean Archipelago

Albers equal area projection
Standard parallels 33° N and 36° N

CITING THIS MAP: Griffith, G.E., Omernik, J.M., McGraw, M.M., Jacobi, G.Z., Canavan, C.M., Schrader, T.S., Mercer, D., Hill, R., and Moran, B.C., 2006, Ecoregions of New Mexico (color poster with map, descriptive text, summary tables, and photographs): Reston, Virginia, U.S. Geological Survey (map scale 1:1,400,000).

Ecoregion maps, publications, GIS files, and contact information are available at www.epa.gov/wed/pages/ecoregions.htm

Level III ecoregion
County boundary
State boundary
International boundary

Acknowledgments

This book was inspired by **Southwestern Trees: A Guide to the Native Species of New Mexico and Arizona** by Elbert L. Little, Jr. (published as Agriculture Handbook No. 9, U.S. Department of Agriculture, December 1950). I came across this small book on my first trip to the Southwest nearly 50 years ago, and put it to use identifying alligator juniper with its distinctive bark. This was to be the first of many memorable trips to the deserts and mountains of this region, and eventually a career studying the flora of this and other regions of the country. Photographs were obtained, where possible, from public domain sources, from the author's own collection, and from a number of photographers on Flickr who have made their work available under Creative Commons commercial use licences (see *www.flickr.com*). Special thanks go to photographers Matt Lavin and Andrey Zharkikh, and the Western New Mexico University Department of Natural Sciences and the Dale A. Zimmerman Herbarium. Detailed plant illustrations were generated from herbarium specimens housed at a number of herbaria in the United States. Grateful acknowledgment is given to the Biota of North America Project (*www.bonap.org*) for permission to use their data to generate the distribution maps.

Online Resources

The following websites offer a wealth of information on the flora of the Southwest:

Flora of North America http://efloras.org
Lady Bird Johnson Wildflower Center https://www.wildflower.org/plants
Oaks of the World http://oaks.of.the.world.free.fr
SEINet http://swbiodiversity.org/seinet
Southwest Desert Flora http://southwestdesertflora.com
The Gymnosperm Database https://www.conifers.org
The Jepson eFlora http://ucjeps.berkeley.edu/eflora/
The Trees of North America http://northamericantrees.com
Vascular Plants of the Gila Wilderness
    https://wnmu.edu/academic/nspages/gilaflora/

Maps
National Individual Tree Species Atlas. 2015, FHTET-15-01. Fort Collins, CO: U.S. Department of Agriculture, Forest Service, Forest Health Technology Enterprise Team.
Kartesz, J.T. 2014. *Floristic Synthesis of North America, Version 1.* Biota of North America Program (BONAP, http://www.bonap.org/).

Index – Scientific Names
Synonyms are listed in *italics.*

Abies, 34
  arizonica, 36
  bifolia, 36
  concolor, 34
  lasiocarpa, 36
Acacia farnesiana, 116
Acacia greggii, 114
Acer, 218
  glabrum, 218
  grandidentatum, 219
  negundo, 221
  neomexicanum, 218
Adoxaceae, 64
AGAVES & CACTI, 231
Ailanthus altissima, 227
Ailanthus Family, 226
Alnus, 68
  incana, 68
  oblongifolia, 70
  tenuifolia, 68
Amelanchier, 176
  australis, 176
  goldmanii, 176
  mormonica, 176
  rubescens, 176
  utahensis, 176
Anacardiaceae, 65
Arbutus, 83
  arizonica, 83
  texana, 85
  xalapensis, *83,* 85
Arecaceae, 253
Asparagaceae, 231
Asparagus Family, 231
Beech Family, 118
Betula, 71
  fontinalis, 71
  occidentalis, 71
Betulaceae, 68
Bignonia Family, 73
Bignoniaceae, 73
Birch Family, 68
Bittersweet Family, 80
BROADLEAF TREES, 64
Broussonetia secundiflora, 91
Buckthorn Family, 166

Bumelia, 224
  lanuginosa, 224
  rigida, 224
Bursera, 75
  fagaroides, 75
  microphylla, 77
  odorata, 75
Bursera Family, 75
Burseraceae, 75
Butcher's-Broom Family, 192
Cactaceae, 240
Cactus Family, 240
Caesalpinia gilliesii, 94
Callitropsis, 17
  arizonica, 18
  glabra, 20
Cannabaceae, 78
Canotia holacantha, 80
Carnegiea gigantea, 240
Cashew Family, 65
Castela emoryi, 226
Celastraceae, 80
Celtis, 78
  laevigata, 78
  reticulata, 78
Cercidium
  floridum, 101
  microphyllum, 102
  torreyanum, 101
Cercis, 89
  arizonica, 89
  occidentalis, 89
Cercocarpus, 178
  betuloides, 178
  breviflorus, 179
  ledifolius, 180
  montanus, 178, 179
  paucidentatus, 179
Cereus
  giganteus, 240
  schottii, 250
  thurberi, 251
Chilopsis linearis, 73
Condalia
  globosa, 166
  lycioides, 174
  obtusifolia, 174

CONIFERS, 17
Cornaceae, 82
Cornus, 82
  alba, 82
  sericea, 82
Cowania, 188
  mexicana, 188
  stansburyana, 188
Crataegus, 181
  cerronis, 181
  chrysocarpa, 181
  erythropoda, 181
  rivularis, 182
  wootoniana, 181
Crown-of-Thorns Family, 148
Cupressaceae, 17
Cupressus
  arizonica, 18
  glabra, 20
Cylindropuntia, 243
  acanthocarpa, 243
  fulgida, 245
  spinosior, 247
  versicolor, 248
Cypress Family, 17
Dalea spinosa, 110
Dermatophyllum secundiflorum, 91
Dogwood Family, 82
Ericaceae, 83
Erythrina flabelliformis, 92
Erythrostemon gilliesii, 94
Euphorbiaceae, 86
Eysenhardtia, 95
  orthocarpa, 95
  polystachya, 95
Fabaceae, 88
Fagaceae, 118
Forestiera, 153
  phillyreoides, 153
  shrevei, 153
Frangula
  betulifolia, 168
  californica, 170
Fraxinus, 154
  anomala, 154
  anomala var. lowellii, 156
  cuspidata, 158
  greggii, 160
  lowellii, 156
  papillosa, 161
  pennsylvanica, 161

  standleyi, 162
  velutina, 162
Fremontia californica, 150
Fremontodendron californicum, 150
Gleditsia triacanthos, 96
Heath Family, 83
Hemp Family, 78
Hesperocyparis
  arizonica, 18
  glabra, 20
Holacantha emoryi, 226
Juglandaceae, 144
Juglans, 144
  major, 144
  microcarpa, 146
  rupestris, 144, 146
Juniperus, 21
  arizonica, 22
  californica, 22
  cedrosiana, 22
  coahuilensis, 22
  deppeana, 24
  erythrocarpa, 30
  monosperma, 26, *30*
  occidentalis, 26
  osteosperma, 28
  pachyphloea, 24
  pinchotii, 30
  scopulorum, 32
  utahensis, 28
Koeberlinia spinosa, 148
Koeberliniaceae, 148
Lemaireocereus thurberi, 251
Lophocereus schottii, 250
Lysiloma, 97
  microphylla, 97
  thornberi, 97
  watsonii, 97
Mallow Family, 150
Malvaceae, 150
Moraceae, 152
Morus microphylla, 152
Mulberry Family, 152
Muskroot Family, 64

Nicotiana glauca, 228
Nightshade Family, 228
Nolina, 192
  bigelovii, 192
  parryi, 192
Oleaceae, 153

Olive Family, 153
Olneya tesota, 98
Opuntia
    acanthocarpa, 243
    fulgida, 245
    spinosior, 247
    versicolor, 248
Ostrya, 72
    baileyi, 72
    knowltoni, 72
Pachycereus schottii, 250
Padus valida, 186
Palm Family, 253
PALMS, 253
Parkinsonia, 99
    aculeata, 99
    florida, 101
    microphylla, 102
    torreyana, 101
Pea Family, 88
Picea, 38
    engelmannii, 38
    parryana, 40
    pungens, 40
Pinaceae, 34
Pine Family, 34
Pinus, 42
    apacheca, 50
    aristata, 43
    arizonica, 45
    cembroides, 46, *47, 48, 56*
    chihuahuana, 54
    edulis, 48
    engelmannii, 50
    flexilis, 52, *60*
    latifolia, 50
    leiophylla, 54
    monophylla, 56
    ponderosa, *45,* 58
    strobiformis, 60
Platanaceae, 164
Platanus, 164
    racemosa, 164
    wrightii, 164
Poinciana gilliesii, 94
Populus, 197
    acuminata, 198
    deltoides subsp. monilifera, 200
    deltoides subsp. wislizeni, 200
    fremontii, *200,* 202
    sargentii, 200

tremuloides, 204
wislizeni, 200
x angustifolia, 198
Prosopis, 104
    glandulosa, 104
    juliflora, 104, 108
    odorata, 106
    pubescens, 106
    velutina, 108
Prunus, 183
    americana, 183
    emarginata, 184
    rufula, 185
    serotina var. rufula, 185
    virens, 185
    virginiana, 186
Pseudotsuga, 62
    menziesii, 62
    taxifolia, 62
Psorothamnus spinosus, 110
Ptelea, 194
    angustifolia, 194
    pallida, 195
    trifoliata subsp. pallida, 195
    trifoliata var. angustifolia, 194
Purshia stansburyana, 188
Quercus, 118
    ajoensis, 142
    arizonica, 120
    chrysolepis, 122, *136*
    diversicolor, 139
    dumosa, 141
    emoryi, 124
    gambelii, 126
    grisea, 128
    gunnisonii, 126
    havardii, 129
    hypoleuca, 130
    hypoleucoides, 130
    leptophylla, 126
    mohriana, 132
    muehlenbergii, 133
    muhlenbergii, 133
    novomexicana, 126
    oblongifolia, 134
    palmeri, 136
    pungens, 138
    reticulata, 139
    rugosa, 139
    submollis, 126
    subturbinella, 141

toumeyi, 140
turbinella, 141, *142*
utahensis, 126
Rhamnaceae, 166
Rhamnus, 168
  betulifolia, 168
  californica, 170
  crocea, 172
  ilicifolia, 172
  ursina, 170
Rhus, 65
  choriophylla, 67
  kearneyi, 65
  ovata, 66
  virens, 67
Ricinus communis, 86
Robinia neomexicana, 112
Rosaceae, 175
Rose Family, 175
Rue Family, 194
Ruscaceae, 192
Rutaceae, 194
Salicaceae, 197
Salix, 206
  amygdaloides, 207
  bebbiana, 208
  bonplandiana, 209
  exifolia, 217
  exigua, 210
  gooddingii, 212
  interior, 211
  laevigata, 213
  lasiandra, 214
  lasiolepis, 215
  microphylla, 217
  nigra, 212
  scouleriana, 216
  taxifolia, 217
  wrightii, 207
Sambucus, 64
  cerulea, 64
  glauca, 64
  mexicana, 64
  nigra, 64
Sapindaceae, 218
Sapindus, 222
  drummondi, 222
  saponaria, 222

Sapium biloculare, 87
Sapotaceae, 224
Sapote Family, 224
Sarcomphalus obtusifolius, 174
Sebastiania biloculares, 87
Senegalia greggii, 114
Sideroxylon lanuginosum, 224
Simaroubaceae, 226
Soapberry Family, 218
Solanaceae, 228
Sophora secundiflora, 91
Spurge Family, 86
Stenocereus thurberi, 251
Strombocarpa, 106
  odorata, 106
  pubescens, 106
Sycamore Family, 164
Tamaricaceae, 229
Tamarisk Family, 229
Tamarix, 229
  aphylla, 230
  gallica, 229
Ungnadia speciosa, 222
Vachellia farnesiana, 116
Vauquelinia californica, 190

Walnut Family, 144
Washingtonia, 253
  robusta, 253
  filamentosa, 253
  filifera, 253
Willow Family, 197
Yucca, 231
  baccata, 239
  brevifolia, 231
  crassifila, 239
  elata, 234
  macrocarpa, 239
  mohavensis, 236
  schidigera, 236
  schottii, 238
  torreyi, 239
  treculeana, 239
Ziziphus obtusifolia, 174
Ziziphus lycioides, 174

Index – Common Names

Acequia Willow, 210
AGAVES & CACTI, 231
Ailanthus, 227
Ailanthus Family, 226
Ajo Mountain Oak, 142
Alamo, 200, 202
ALDER, 68
Alligator Juniper, 24
Allthorn, 148
Almond Willow, 207
Alpine Fir, 36
American Plum, 183
Apache Pine, 50
Arizona Alder, 70
Arizona Ash, 162
Arizona Black Walnut, 144
Arizona Blueberry Elder, 64
Arizona Cypress, 18
Arizona Elder, 64
Arizona Longleaf Pine, 50
Arizona Madrone, 83
Arizona Madrono, 83
Arizona Oak, 120
Arizona Pine, 45
Arizona Planetree, 164
Arizona Ponderosa Pine, 45
Arizona Redbud, 89
Arizona Rosewood, 190
Arizona Sycamore, 164
Arizona Walnut, 144
Arizona White Oak, 120
Arizona Yellow Pine, 45
Arroyo Willow, 215
ASH, 154
Asparagus Family, 231
Aspen, 204
Athel Tamarisk, 230

Balsam Fir, 34
Basket Willow, 210
Beaked Willow, 208
Bebb's Willow, 208
Beech Family, 118
Bellota, 124
Bigelow Nolina, 192
Bigelow's Beargrass, 192

Bignonia Family, 73
Bigtooth Maple, 219
Birch Family, 68
Birchleaf Buckthorn, 168
Birchleaf Mountain-Mahogany, 178
Bird-Of-Paradise Shrub, 94
Bitter Cherry, 184
Bitter Condalia, 166
Bittersweet Family, 80
Black Birch, 71
Black Cottonwood, 198
Black Haw, 182
Black Oak, 124
Black Willow, 216
Blackjack Oak, 124
Blue Douglas-Fir, 62
Blue Paloverde, 101
Blue Spruce, 40
Bonpland Willow, 209
Border Pinyon, 47
Border Limber Pine, 60
Boxelder, 221
Bristlecone Pine, 43
BROADLEAF TREES, 64
Buckhorn Cholla, 243
Buckthorn Family, 166
Buckthorn, 168
Bursera Family, 75
BURSERA, 75
Butcher's-Broom Family, 192

Cactus Family, 240
California Buckthorn, 170
California Fan-Palm, 253
California Fremontia, 150
California Juniper, 22
California Redbud, 89
California Scrub Oak, 141
California Slippery-Elm, 150
California Washingtonia, 253
California-Palm, 253
Cane Cholla, 247
Canotia, 80
Canyon Live Oak, 122, 136
Cashew Family, 65
Castor-Bean, 86

Catclaw Acacia, 114
Catclaw, 114
Cerro Hawthorn, 181
Chainfruit Cholla, 245
Cherioni, 222
CHERRY, 183
Chihuahuan Pine, 54
Chihuahuan Ash, 161
Chilicote, 92
Chinquapin Oak, 133
Chisos Wild Cherry, 185
Chittamwood, 224
Cholla, 243
Coffeeberry, 170
Colorado Blue Spruce, 40
Colorado Pinyon Pine, 48
Colorado Spruce, 40
Common Chokecherry, 186
CONIFERS, 17
Copal, 77
Coralbean, 91
Corkbark Fir, 36
Corona de Cristo, 148, 226
Cottonwood, 197
Coyote Willow, 210
Crown-of-Thorns Family, 148
Crown-of-Thorns, 148
Crucifixion-Thorn, 80, 148, 226
Curlleaf Mountain-Mahogany, 180
CYLINDROPUNTIA, 243
Cypress Family, 17
CYPRESS, 17

Datil Yucca, 239
Deerhorn Cholla, 248
Desert Ash, 162
Desert Elderberry, 64
Desert Ironwood, 98
Desert-Olive Forestiera, 153
Desert-Olive, 153
Desert-Willow, 73
Desertpalm, 253
Devilsclaw, 114
Dogwood Family, 82
DOUGLAS-FIR, 62
Douglas-Spruce, 62
Dudley Willow, 212
Dwarf Maple, 218

Eastern Mountain-Mahogany, 179
Elephant Bursera, 77

Elephant-Tree, 77
Emory Oak, 124
Engelmann Spruce, 38

FIR, 34
Fireberry Hawthorn, 181
Fire Willow, 216
Flannelbush, 150
Flowering Ash, 158
Foothill Paloverde, 102
Foxtail Pine, 43
Fragrant Ash, 158
Fragrant Bursera, 75
Fremont Cottonwood, 202
Fremont Poplar, 202
Fremont Screwbean, 106
French Tamarisk, 229
Fresno, 162
Frijolito, 91

Gambel Oak, 126
Giant Cactus, 240
Gila Chokecherry, 185
Goat-Bean, 91
Golden Aspen, 204
Goodding's Willow, 212
Gray Oak, 128
Gray-Thorn, 174
Green Ash, 161
Gregg's Ash, 160
Gum Bumelia, 224
Gum-Elastic, 224

Hairy Mountain-Mahogany, 179
Havard's Oak, 129
HAWTHORN, 181
Heath Family, 83
Hemp Family, 78
Hoary Yucca, 238
Holacantha, 226
Hollyleaf Buckthorn, 172
Hollyleaf Redberry Buckthorn, 172
Honey Mesquite, 104
Honey-Locust, 96
Hoptree, 194
Horsebean, 99
Huisache, 116

Indian-Bean, 92
Indigobush, 110
Inland Boxelder, 221

Jaboncillo, 222
Jerusalem-Thorn, 99
Joshua-Tree Yucca, 231
Joshua-Tree, 231
Judas-Tree, 89
Jumping Cholla, 245
Jumping-Bean Sapium, 87
Junco, 148
JUNIPER, 21
  Alligator, 24
  California, 22
  One-Seed, 26
  Pinchot, 30
  Redberry, 22
  Rocky Mountain, 32
  Utah, 28

Kearney's Sumac, 65
Kidneywood, 95
Knowlton's Hop-Hornbeam, 72

Lanceleaf Cotonwood, 198
Limber Pine, 52
Little Walnut, 146
Littleleaf Ash, 160
Littleleaf Horsebean, 102
Littleleaf Lysiloma, 97
Littleleaf Paloverde, 102
Lotebush, 174
Lotewood Condalia, 174
Lowell Ash, 156

MADRONE, 83
Mallow Family, 150
Manzana de Puya Larga, 181
MAPLE, 218
Mearns' Sumac, 67
Mescalbean Sophora, 91
Mescalbean, 91
Mescat Acacia, 116
MESQUITE, 104
Mexican-Buckeye, 222
Mexican Blue Oak, 134
Mexican Cliffrose, 188
Mexican Elder, 64
Mexican Jumping-Bean, 87
Mexican Mulberry, 152
Mexican Paloverde, 99
Mexican Pinyon Pine, 46
Mexican Pinyon, 46
Mexican Washingtonia, 253

Mexican White Pine, 60
Mohave Yucca, 236
Mohr's Oak, 132
Mohr's Shinoak, 132
Mountain Alder, 68
Mountain Cottonwood, 198
Mountain Leatherwood, 150
Mountain Mulberry, 152
Mountain Spruce, 38
Mountain Willow, 216
Mountain Yucca, 238
Mountain-Laurel, 66
MOUNTAIN-MAHOGANY, 178
Mulberry Family, 152
Muskroot Family, 64

Narrowleaf Cottonwood, 198
Narrowleaf Hoptree, 194
Narrowleaf Poplar, 198
Netleaf Hackberry, 78
Netleaf Oak, 139
New Mexican Alder, 70
New Mexico Locust, 112
Nightshade Family, 228
Nogal, 144, 146
Nut Pine, 46, 48, 56

OAK, 118
  Ajo Mountain, 142
  Arizona, 120
  Arizona White, 120
  Black, 124
  Blackjack, 124
  Canyon Live, 122, 136
  Chinquapin, 133
  Emory, 124
  Gambel, 126
  Gray, 128
  Havard's, 129
  Mexican Blue, 134
  Mohr's, 132
  Mohr's Shinoak, 132
  Netleaf, 139
  Palmer, 136
  Pungent, 138
  Rocky Mountain White, 126
  Sandpaper, 138
  Scrub, 141
  Shinnery, 129
  Shrub Live, 141
  Silverleaf, 130

Toumey's, 140
Turbinella, 141
Utah White, 126
Whiteleaf, 130
Olive Family, 153
One-Seed Juniper, 26
Oregon Pine, 62
Organpipe Cactus, 251

Pacific Willow, 214
Pale Hoptree, 195
Palm Family, 253
Palma, 239
Palmer Oak, 136
PALMS, 253
Palo de Hierro, 98
Palo Fierro, 98
Paloblanco, 78
PALOVERDE, 99
Pea Family, 88
Peach Willow, 207
Peachleaf Willow, 207
Pigeonberry, 170
Pinchot Juniper, 30
Pine Family, 34
PINE, 42
    Apache, 50
    Arizona, 45
    Arizona Yellow, 45
    Border Limber, 60
    Bristlecone, 43
    Chihuahuan, 54
    Colorado Pinyon, 48
    Foxtail, 43
    Limber, 52
    Mexican Pinyon, 46
    Mexican White, 60
    Nut, 46, 48, 56
    Oregon, 62
    Pinyon, 46, 48
    Rocky Mountain White, 52
    Singleleaf Pinyon, 56
    Southwestern White, 60
    Western Yellow, 58
    White, 52
    Yellow, 58
Pitahaya Dulce, 251
Pitahaya, 251
Planetree, 164
Plains Cottonwood, 200

Plains Poplar, 200
Polished Willow, 213
Ponderosa Pine, 58
PTELEA, 194
Pungent Oak, 138

Quaking Aspen, 204
Quinine-Bush, 188

Red Birch, 71
Red Fir, 62
Red Willow, 213
Red-Osier Dogwood, 82
Redberry Juniper, 22
Retama, 99
Rio Grande Cottonwood, 200
Rio Grande Poplar, 200
River Hawthorn, 182
Rocky Mountain Boxelder, 221
Rocky Mountain Juniper, 32
Rocky Mountain Maple, 218
Rocky Mountain Redcedar, 32
Rocky Mountain White Oak, 126
Rocky Mountain White Pine, 52
Rose Family, 175
Rough-Bark Arizona Cypress, 18
Rue Family, 194

Saguaro, 240
Salt-Cedar, 229
Sandbar Willow, 210, 211
Sandpaper Oak, 138
Sapote Family, 224
Sauco, 64
Schott's Yucca, 238
Scouler's Willow, 216
Screwbean Mesquite, 106
Scrub Oak, 141
Senita, 250
Shinnery Oak, 129
Shrub Live Oak, 141
Silver Fir, 34
Silver Spruce, 38, 40
Silverleaf Oak, 130
Singleleaf Ash, 154
Singleleaf Pinyon Pine, 56
Singleleaf Pinyon, 56
Smokethorn Dalea, 110
Smokethorn, 110
Smoketree, 110

Smooth Arizona Cypress, 20
Smooth Ash, 162
Soapberry Family, 218
Soaptree Yucca, 234
Soapweed, 234
Southwestern Chokecherry, 185
Southwestern Coralbean, 92
Southwestern Locust, 112
Southwestern Peach Willow, 207
Southwestern White Pine, 60
Spanish Bayonet, 238, 239
Spanish Dagger, 236, 238, 239
Speckled Alder, 68
SPRUCE, 38
Spurge Family, 86
Staghorn Cholla, 248
Subalpine Fir, 36
Sugar Sumac, 66
Sugarbush, 66
SUMAC, 65
Sweet Acacia, 116
Sycamore Family, 164

Tamarisk, 229
Tamarisk Family, 229
Tasajo, 247
Tesota, 98
Texas Black Walnut, 146
Texas Madrone, 85
Texas Madrono, 85
Texas Mulberry, 152
Thinleaf Alder, 68
Tornillo, 106
Torote, 77
Torrey Vauquelinia, 190
Torrey Yucca, 239
Toumey Ash, 162
Toumey Willow, 209
Toumey's Oak, 140
Tree-of-Heaven, 227
Tree Cholla, 248
Tree Tobacco, 228
Trembling Poplar, 204
Turbinella Oak, 141
Two-Needle Pinyon, 48

Una De Gato, 114
Utah Juniper, 28

Utah Serviceberry, 176
Utah White Oak, 126

Valley Cottonwood, 200
Velvet Ash, 162
Velvet Mesquite, 108

Walnut Family, 144
WALNUT, 144
Water Birch, 71
Western Black Willow, 212, 214
Western Coralbean, 92
Western Hackberry, 78
Western Hophornbeam, 72
Western Juniper, 24, 28, 32
Western Redbud, 89
Western Soapberry, 222
Western Sugar Maple, 219
Western Yellow Pine, 58
Whiplash Willow, 214
White Balsam, 34
White Balsam, 36
White Crucillo, 174
White Fir, 34, 36
White Pine, 52
White Spruce, 38
White Willow, 215
Whiteleaf Oak, 130
Wild Cherry, 184, 186
Wild China-Tree, 222
Wild Plum, 183
Wild-Olive, 153
White-Thorn, 116
Willow Family, 197
WILLOW, 206
Wislizenus Cottonwood, 200
Wooten's Hawthorn, 181
Wright Mountain-
Mahogany, 179

Yellow Paloverde, 102
Yellow Pine, 58
Yellow Willow, 214
Yew Willow, 217
Yew-Leaf Willow, 217
YUCCA, 231

www.ingramcontent.com/pod-product-compliance
Lightning Source LLC
Chambersburg PA
CBHW051243020426
42333CB00025B/3033